Multi-modality Control of Reconfigurable Swarm

Tao Jiang · Zhi Zheng · Lan Yao · Jianchuan Ye ·
Jianxiang Wang · Jingxiang Feng

Multi-modality Control of Reconfigurable Swarm

Springer

Tao Jiang
Chongqing University
Chongqing, China

Lan Yao
Chongqing Vocational College
of Transportation
Chongqing, China

Jianxiang Wang
Chongqing University
Chongqing, China

Zhi Zheng
Jiangsu Automation Research Institute
Lianyungang, China

Jianchuan Ye
Beijing Institute of Technology
Beijing, China

Jingxiang Feng
Jiangsu Automation Research Institute
Lianyungang, China

ISBN 978-981-95-0304-9 ISBN 978-981-95-0305-6 (eBook)
https://doi.org/10.1007/978-981-95-0305-6

This work was supported by National Natural Science Foundation of China (62106027) and National Key Research and Development Program of China Stem Cell and Translational Research (2019YFB1312000).

This Springer imprint is published by the registered company Springer Nature Singapore Pte Ltd.
The registered company address is: 152 Beach Road, #21-01/04 Gateway East, Singapore 189721, Singapore

Preface

With the rapid advancement of unmanned and intelligent technologies, mobile robots are accelerating their application scenarios from structured zones to complex, diverse zones. Unlike structured environments such as ports, factories, and households, unstructured scenarios like abandoned buildings and wild caves exhibit heterogeneous terrain variations and narrow interlaced passages. Existing vehicles with fixed configurations or weakly reconfigurable designs demonstrate limited flexibility in capability deployment, struggling to balance "capability" and "efficiency". As a novel swarm system, reconfigurable swarms consist of multiple modular units capable of both independent mobility and collective reconfiguration. Through physical connections between modular units, the system adaptively reconfigures into combination modalities of varying scales to move complex terrains, while decomposing into distributed wireless-networked units via virtual connections for efficient parallel operations in flat or confined spaces. Compared with traditional swarms and large fixed-configuration vehicles, reconfigurable swarms achieve on-demand capability aggregation and dispersion in response to environmental features and mission requirements, demonstrating superior adaptability in unstructured scenarios. This positions them as a critical research direction for overcoming application bottlenecks in robotic operations within complex unstructured environments.

Reconfigurable swarms achieve scenario adaptability through modal switching between modular units and combination modalities, along with multi-modality behavior control during group operations. However, compared to conventional swarm systems and reconfigurable swarm systems, they require coordinated operation across three modalities: individual units, combination modality, and swarms. Significant differences exist in task requirements and response mechanisms across modalities, resulting in higher-dimensional system variables and more complex operational scenarios. Moreover, individual unit modality control involves more constraints and intensified uncertain disturbances compared to homogeneous vehicle control. Multimodal Reconfiguration control further presents challenges, including real-time modality-scenario matching, optimal unit selection, and precise rapid docking, making technical implementation particularly demanding. This monograph focuses on reconfigurable swarms, researching autonomous swarm control to establish a comprehensive technical framework for adaptive multi-modular swarm control.

The technical exposition follows a progressive architecture of "theoretical foundations-core technologies-system integration-application validation", encompassing reconfigurable swarm multimodal modeling, coordination control, combination modality control, reconfiguration docking, and modal scheduling, essentially covering all core procedures of autonomous swarm control. The book also details the design methodology of a self-developed physical platform for reconfigurable swarms, with all theoretical approaches systematically validated through platform-based experiments.

The monograph comprises eight chapters organized as follows:

Chapter 1: Analyzes mission requirements and technological trends, elucidating the strategic value of reconfigurable swarms in disaster rescue and wilderness exploration scenarios.

Chapter 2: Elaborates on hardware design and perception-communication schemes for reconfigurable swarm, with experimental validation of mobility functions and platform construction of reconfigurable coordination.

Chapter 3: Establishes kinematic-dynamic models for modular unit and combination modality, developing a kinematics-dynamics cascaded controller based on low-peak extended state observers.

Chapters 4 and 5: Address swarm control under nonholonomic constraints, enhancing control system performance through saturation oscillation suppression and distributed disturbance observers.

Chapter 6: Proposes a dynamic reconfiguration control framework resolving docking stability and energy optimization in large-scale reconfiguration scenarios.

Chapter 7: Develops an anti-disturbance and fault-tolerant framework for combination modalities, focusing on fault diagnosis and redundant actuation optimization.

Chapter 8: Designs a task-environment dual-driven multimodal scheduling strategy enabling scenario-adaptive and autonomous modal transitions.

All technical components have been verified through high-fidelity simulations and physical experiments, demonstrating algorithm effectiveness. This systematic investigation into multi-modular autonomous control methods for reconfigurable swarms not only provides theoretical insights for academic control research but also offers practical references for addressing robotic operations in unstructured environments within engineering domains.

Chongqing, China Tao Jiang
May 2025

Acknowledgements The journey from the inception of reconfigurable swarm research to the finalization of this monograph spanned 4 years. All research activities were conducted under the organizational framework of the Laboratory of Robotics and Intelligent Systems at Chongqing University, which provided robust research infrastructure and spatial resources to ensure the progression of this work. We hereby express our profound gratitude to the laboratory for its indispensable support.

Special gratitude is extended to Prof. Su Xiaojie, the academic mentor of this monograph. His critical feedback and suggestions, consistently grounded in visionary academic acumen and a rigorous scientific approach, have guided the research direction with exceptional clarity.

The collective dedication of the research team served as the cornerstone for the completion of this work. We extend heartfelt appreciation to Lei Du, Chen Wang, Yang Xiao, Mingxing Dou, Zeyu Xie, Hao Zhang, Rongqin Mo, and Zuming Xiong for their diligent contributions to platform development, experimental validation, and data analysis. Their relentless efforts were pivotal to the smooth execution of this research and the successful compilation of this monograph.

Finally, we dedicate this monograph to all colleagues committed to advancing intelligent technologies in mobile robotics. May our modest explorations infuse new vitality into the future development of unmanned systems.

Competing Interests The authors have no competing interests to declare that are relevant to the content of this manuscript.

Contents

Chapter 1
Introduction

Abstract The technological and application advantages of reconfigurable swarms over traditional swarms are illustrated in this chapter, and a comprehensive and in-depth discussion is presented on the mission and technological overview, future development trends, and technical directions of this new type of swarm, as well as its strategic value in scenarios such as disaster relief and field exploration.

Keywords Reconfigurable swarms · Mission and technological overview · Future development trends · Technical directions

1.1 Background

In recent years, with the rapid development of automation and intelligent technologies, the application scenarios of mobile intelligent entities have evolved from simple structured environments to increasingly complex unstructured environments. China's "Fourteenth Five-Year Plan for Robot Industry Development", the United States' "Robot Plan 2.0", Japan's "New Robot Strategy", and other top-level designs for the development of mobile intelligent entities in different countries have clearly proposed to accelerate the deep application of mobile intelligent entities in complex scenarios such as urban search and rescue, mine exploration, underground detection, etc. However, the existing mobile swarms still have the pain point of difficulty in balancing capability and efficiency. For example, in urban search and rescue scenarios, small mobile intelligent entities can execute the search for survivors in vacant areas with high efficiency, but they struggle to pass through obstacle terrains such as stairs and trenches; large mobile intelligent entities have concentrated capabilities and strong passability, but due to their large size and high cost, they are not suitable for executing tasks in a dispersed manner in complex urban environments or responding quickly to emergencies. Therefore, it is urgent to develop a mobile swarm system that can disassemble and assemble according to time and demand, balancing capability and efficiency. The reconfigurable swarm with distributed execution and organization reconfiguration characteristics is an ideal choice to solve the above pain points.

T. Jiang et al., *Multi-modality Control of Reconfigurable Swarm*,
https://doi.org/10.1007/978-981-95-0305-6_1

Reconfigurable swarms are a new type of multi-agent system composed of small modular units that can move independently and reconfigure freely. Compared to traditional mobile swarms that are constrained by the efficiency and capability when executing complex scenario tasks, this new type of swarm system can converge and decompose capabilities in a timely and as-needed manner according to complex terrains and environmental situations. For example, during urban search and rescue, on one hand, large-scale reconfigurable small modular swarms can be quickly brought to the disaster site by rescue personnel, and the small units can be controlled to form multiple points scattered across a wide area to carry out search tasks, responding swiftly to the disaster. On the other hand, multiple units can be used to reconfigure into various combined structures, flexibly traversing stairs, ditches and ruins. After overcoming obstacle terrains, they can timely decompose into a multi-unit cooperative mode; at the same time, when a companion unit fails, it can be assisted by the redundant actuation after the convergence and reconfiguration of healthy units to restore its mobility. In summary, reconfigurable swarms use the reconfiguration expansion when small units converge and the swarm enhancement when they disperse, achieving a balance between the operational capability and efficiency of mobile swarms. Due to their late emergence and novel configuration, existing results are mostly focused on structural design and capability testing, and there is still a lack of a control theory system and application framework to support their practical application. In terms of multi-modal control, there are two key scientific problems of "modality transition" and "performance improvement" under complex environments, specifically as follows:

(1) Reconfigurable swarms exist in various forms, including modular units and combined modalities of different reconfiguration scales. They exhibit multi-modal characteristics interwoven between dispersed execution and aggregated reconfiguration. Designing to promote the modality switching and multi-modal behavior coordination of individual units and combined modalities to ensure the smooth transition of the mobile swarm in complex scenarios is very challenging (smooth modality transition).

(2) The task execution environment of reconfigurable swarms is quite harsh. The control of its individual units and combined modalities in practical use not only easily triggers saturation, safety and other multi-dimensional restrictive constraints, but also continuously suffers from model uncertainty, external load changes, random actuation failures and other multi-source uncertainties. Designing the control of individual units and combined modalities to enhance adaptability under complex environmental constraints and disturbances is extremely challenging (control adaptability enhancement).

In summary, combining the current research background and the two key scientific problems limiting the application of reconfigurable swarms, this project specifically carries out research on multi-modal control of reconfigurable swarms adaptable to complex scenarios, truly forming a new system of mobile swarm operations with dual efficiency in complex scenarios.

1.2 Reconfigurable Swarm Review

This section provides an overview of the design, modeling and control, reconfiguration docking and integrated coordination of reconfigurable swarms. Overall, the research progress is mainly focused on the design of the ontology mechanism and hardware architecture and the capability verification carried out around its configuration adaptability, with less attention paid to research results at other levels.

1.2.1 Reconfigurable Swarm Ontology Research

The current main use of reconfigurable swarm carriers is to utilize high-efficiency mobile mechanisms such as wheels and tracks, endowing individual units with independent mobility, and completing the reconfiguration transformation between individual units/combination modalities through mechanical connections, electromagnetic coupling, and self-healing alloys and other technical means, generating new configurations that adapt to the current tasks/terrain.

The ModLab team at the University of Pennsylvania in the United States has developed a reconfigurable SMORES-EP, which uses wheel differential drive to achieve flexible movement of the modular. Researchers at the Polytechnic University of Madrid in Spain designed a wheel reconfigurable swarm RobMAT and a single-point interlocking connection device, and tested the reconfiguration capabilities of multi-unit chain combinations [1]. Luo Hongxing et al. developed the Sambot series of wheel reconfigurable swarms, which mainly consists of differential drive and docking panels, and can actively dock to reconfigure into various chain/lattice combination modalities on all four sides [2]. Similar multi-face docking includes the M2SBot wheel reconfigurable swarm designed by Ge et al., which can be reconfigured into line/surface/body rich combination structures through the plug-in structure [3]. The Omni-Pi-tent self-assembly and self-recovery modular swarm, designed by the Safe Autonomous Research Institute at the University of York in the UK, features omnidirectional drive of Mecanum wheels, four-sided genderless mechanical docking, and the ability to actively reconfigure into chain or three-dimensional lattice combination structures [4]. Zhou et al. developed a reconfigurable swarm with omnidirectional flexible drive of Mecanum wheels, and installed magnetic devices on its side and transverse surfaces to support line/surface combination structure reconfiguration [5], to achieve rich configurations in coordination. Similar wheel-driven reconfigurable swarms include MIRRAX designed by the research team at the University of Manchester in the UK [6], and M3Express developed by the Applied Physics Laboratory team at Johns Hopkins University in the United States [7, 8].

In addition to wheel actuation, many researchers have chosen the more complex mechanical actuation but stronger mobility of the caterpillar actuation as the movement mode of the modular unit of the reconfigurable swarm. Wang et al. developed a caterpillar actuation reconfigurable mobile unit JL-2 with a hook claw connection device, and embedded a spherical joint at the connection to achieve rich combination structures to adapt to complex terrain [9, 10]. Liu et al. developed a caterpillar drive reconfigurable swarm MMRP with four docking interfaces, and used a combination cam-spherical connection mechanism, combined with a hybrid reconfiguration transformation planning algorithm based on graph theory, to test the active generation capability of combination structures in swarm tasks [11]. Similar caterpillar-type reconfigurable swarms include Millibots designed by the team at Carnegie Mellon University in the United States, STORM developed by the team at the University of Washington in the United States [12, 13], R-Track developed by researchers at Hanyang University in South Korea [14], and Crawler Unit designed by the Italian Institute of Technology [15]. In recent years, new types of reconfigurable swarms have begun to emerge, such as the Freeform swarm series developed by the team led by Lin et al., including FreeBOT [16], SnailBot [17], and FreeSN [18]. For the first time internationally, they proposed a continuous connection mechanism between modules, further promoting theoretical innovation in the application algorithms of freeform reconfigurable swarms in fluid obstacle crossing, sampling operations, etc (Fig. 1.1).

1.2.2 Reconfigurable Swarm Modeling and Control Research

Reconfigurable swarm modeling and control can be divided into two major categories based on different structural features: single modular units and combined modalities. Although the modeling and control of single modular unit are the same as those of mobile vehicles with the same mobility, their homomorphic drive models and control methods can be referred to. However, research results related to the modeling and control of combined modalities of any reconfigurable scale are quite rare.

(1) Reconfigurable Swarm Modeling Research

The construction of reconfigurable swarm models is the basis for the design of single unit and combined modality controllers, the promotion of rapid switching of reconfigurable interfaces between modular units and combined modalities, and the design of a multi-modal coordination framework that integrates all control methods. Unit modality modeling refers to the establishment of mathematical models that describe the kinematic and dynamic model characteristics of each independent module unit. This type of model can fully refer to the construction methods of homomorphic mobile vehicle models, such as the nonholonomic constraint model of a two-wheel differential actuation mobile unit [19–22], the omnidirectional motion model of a Mecanum wheel actuation mobile unit [5, 23, 24], and the differential steering model of a tracked drive mobile unit [13, 25, 26]. , and the omnidirectional rolling model of spherical mobile vehicles [27], etc.

The greatest challenge in modeling reconfigurable swarms compared to existing fixed configuration mobile swarms lies in the variability of the combined structure,

Fig. 1.1 Reconfigurable swarm

the scalability of dimensions, and the construction of universality. That is, the actual application model construction method is required to automatically generate matching kinematics and dynamics models for mobile unit swarms of any reconfiguration scale. Lashkari et al. [19] established a kinematic model of dual-unit reconfiguration under non-integral constraints based on structural parameters, but the assumption of trailing faults in the unit after modeling limited the adaptability of larger scale reconfigurations within the swarm. Ercan et al. [28] designed two motion modes, climbing and differential driving, within the ULGEN reconfigurable swarm, and introduced a homogeneous transformation matrix to describe the gait of the climbing mode and the posture of each unit in differential driving. Shi et al. [29] established a combined modality kinematic model that ensures coordinated motion between modules by modeling the instantaneous center of rotation (ICR) space, combined with the independent motion and kinematic constraints of the differential drive module unit.

In addition to the above wheel drive modeling, Li et al. [30] combined Denavit-Hartenberg parameters to establish the four-bar wheel-leg structure and the forward and inverse kinematics models of the active docking end of the wheel-leg reconfigurable swarm; Zong et al. [31] constructed a forward and inverse kinematics model describing the motion relationship between the joint and the end effector of the combined modality by defining the rotation matrix, position vector, and free mode matrix of the spherical rolling contact joint, and combining the no-slip constraint condition.

All of the above only involve the construction of the kinematic model of the reconfigurable swarm combined modality, but this swarm system is mostly applied in complex scenes, and needs to consider internal characteristics such as mass and inertia of the members, and external factors such as load, friction, and disturbance to support the design of high-performance robust combined modality controllers. The modeling of the dynamics of the reconfigurable swarm combined modality is even more scarce compared to the kinematics results. Fernando et al. [32] used a symbiotic strategy to simulate and evaluate the dynamics of the self-reconfigurable swarm three-unit combination and six-wheel model in terms of load, velocity, and other multi-dimensional performance optimization, but lacked details of model construction. Leggieri et al. [33] derived a combined modality kinematic model constrained by the kinematic constraints on the track after recombination and structural parameters, and obtained a dynamic model of velocity and force/torque response based on the Lagrangian method, but lacked the derivation process of the bottom dynamics and the specific form of its model. Shi et al. [34] analyzed the torque requirements of the conveyor belt system and the change of the center of gravity between modules, but still need to describe the dynamic model of the swarm's complex mode overall characteristics to optimize controller design. Existing results have not yet deeply described the high walking efficiency wheel/track drive expandable universal combination modality dynamic modeling method with position-velocity-torque characteristics, but there are theoretical results of dynamic construction of reconfigurable swarms with other motion modes.

In summary, although the current control model for reconfigurable swarms that satisfy the requirements of configurability, dimension scalability, and universal construction is still scarce, it is highly feasible to derive the kinematic and dynamic models of the combined modality under any reconfiguration scale based on general modeling methods and theoretical tools.

(2) Coordinated Control Research on Reconfigurable Swarms

The control research within reconfigurable swarms can also be divided into single unit control and combined modality control. Although the current reconfigurable swarm units mostly adopt open-loop or simple kinematic closed-loop control [24, 35], they can refer to the control methods of isomorphic actuation mobile units, and fully introduce various advanced control methods such as backstepping [36], sliding mode [37], and model predictive control [38], to enhance the control adaptability of reconfigurable swarm units in complex environments.

The control of reconfigurable swarms within combined modalities as a new type of multi-agent system is a key "new" area, with equally scarce research results in control and modeling. Li et al. [39] used experimental research to study the matching relationship between the configuration of the wheel-driven self-reconfigurable combined modalities and terrain parameters. Pfotzer et al. [40] introduced an optimization strategy to control the obstacle-crossing locomotion of the snake-like configuration after reconfiguration, but none of them involved precise control of the kinematic and dynamic state feedback of the combination modality. Shi et al. [29] used a hierarchical control framework based on existing models, but lacked instantaneous center of rotation (ICR) positioning and steering feedback control, and lacked robust response of the bottom dynamics, limiting its practical application in complex environments. Lashkari et al. [19] took advantage of the fault tolerance of redundant actuation in reconfigurable swarm combined modalities, designed a chain transformation model based on the kinematics of a dual-unit combined modality with a rear unit failure, designed a formation coordination and trajectory stabilization controller, and designed an adaptive robust velocity tracking controller at the level of the healthy unit's bottom dynamics, but this cascaded control is limited by the lack of universality of the model.

In summary, the reconfigurable swarm, as a new form of swarm being explored in the laboratory scenario, has the support of isomorphic actuation mobile unit control theory in modeling and control, and a little exploratory theory and application work. However, it is still not enough to support the practical application of the swarm system unit and the combination in complex environments. Therefore, the research on the control of reconfigurable swarm modular units should fully consider the interference of many negative factors such as saturation, safety, model and other various restrictive constraints, internal model uncertainty, and external unknown disturbances in its future operation scenarios, expand various advanced control theories and methods to enhance its adaptability in the process of clustering and reconfiguration; the research on the control of the combination within the swarm should focus on completing the control theory system of the existing reconfigurable swarm as a whole, and under the consideration of the above constraints and disturbances, construct a fault-tolerant, safe, robust and all-round combination controller for any reconfiguration scale, to maximize the advantages of the combination's permeability and adaptability in complex environments.

1.2.3 Reconfiguration Docking and Task Coordination Scheduling of Reconfigurable Swarms

The reconfiguration docking within the reconfigurable swarm realizes the rapid configuration transition of the unit and the combination, which is the key process for its capability enhancement, and the integrated coordination of multi-modal behaviors is the concentrated embodiment of the dual superiority of the reconfigurable swarm in complex scenario tasks.

(1) Reconfiguration Docking of Reconfigurable Swarms
The existing research on the docking of reconfigurable swarms largely depends on closed laboratory scenarios, relying on high-precision sensors for positioning under a single positioning reference (inertial system or vehicle system), sensing the moving unit and the surrounding environment, and driving the modular unit or combination modality to complete the various combination reconfiguration. For example, Huang et al. [41] used infrared transceivers, Tan et al. [42] adopted visual and laser sensing, and Quiñonez et al. [1] used ultrasonic and visual mixed sensing of the environment. The sensing range of the above airborne sensors (vehicle system) is limited, and a random search process is introduced outside the sensing range to determine the direction of the docking target. However, to improve execution efficiency, reconfigurable swarms often have a wide distribution of modular unit or combination modalities, which cannot guarantee that the docking targets are all within the effective sensing range. Although the existing sensing and docking strategies can meet the functional verification requirements of laboratory scenarios, they greatly limit the true performance of reconfigurable swarms and reconfiguration tasks. Wei et al. [43] used high-precision cameras for global positioning (inertial system) to provide pose and environmental information for swarm members, but the sensing accuracy of the actual application of global positioning methods (outdoor navigation and indoor mapping) is usually difficult to meet the mechanical reconfiguration requirements of close-range docking mechanisms. At the same time, current research on docking reconfiguration mainly focuses on docking strategies, and the bottom control mostly uses open-loop or simple kinematic closed-loop control methods [1], not only ignoring the interference of kinematic tracking accuracy, but also not considering the negative impact of many application constraints and lumped disturbances inside and outside the model, such as performance saturation, environmental safety, and limited sensing. In summary, the existing docking strategies and control methods cannot effectively and adaptively promote the reliable switching of units and combination modalities within the swarm.

(2) Coordinated Scheduling of Multimodal Behavior in Reconfigurable Swarms
The coordinated scheduling of multimodal behavior in reconfigurable swarms is a key step in realizing its potential in various applications, and related results are still relatively rare. Lashkari et al. designed a hybrid formation controller for reconfigurable swarm units and reconfigured combination modalities after failure, a docking controller for post-failure reconfiguration, and an obstacle avoidance controller for

unit/combination modalities at the kinematic level. They also used the health status of swarm members to design reconfiguration triggering and formation rebuilding strategies, achieving multimodal behavior scheduling of swarm control and its reconfiguration docking process for unit/combination modalities [19]. Jing et al. abstracted the action behavior, workspace, and environmental events of reconfigurable swarm members into a set of states and events, and used a finite state machine to design a multimodal behavior coordinated scheduling framework, driving the modular units and various combination modalities within the swarm to respond to dynamic environmental events [44]. Overall, the essence of coordinated scheduling is a high-level behavior decision problem for mobile units in reconfigurable swarms based on their own state and external environment. It can be implemented using generic decision methods such as finite state machines and behavior trees to achieve efficient transitions of multimodal hybrid behaviors such as unit swarm, fault reconfiguration, obstacle crossing reconfiguration, combination modality obstacle crossing, and mixed swarm of unit and combination modalities.

1.3 Future Applications and Research Directions of Reconfigurable Swarms

1.3.1 Future Application Prospects of Reconfigurable Swarms

With the rapid development of frontier technologies such as robotics, artificial intelligence, the Internet of Things (IoT), and advanced manufacturing, reconfigurable swarms will show broad application prospects in multiple key areas. Their unique swarm coordination and reconfiguration capabilities make such swarm systems capable of efficiently executing tasks in complex dynamic environments. In the future, the application of reconfigurable swarms will gradually move from laboratory verification to actual deployment, and play a core role in multiple fields such as smart cities, disaster rescue, national defense security, industrial manufacturing, and space exploration.

(1) Smart Cities and Automated Infrastructure Maintenance
Future smart cities rely on highly automated and intelligent infrastructure maintenance systems, and reconfigurable swarms are one of the key technologies to achieve this goal. For example, their high flexibility and reconfigurability can be widely applied to urban inspection and maintenance. Reconfigurable swarms can be deployed in areas difficult for humans to reach, such as urban underground pipe networks, bridges, and exterior facades of high-rise buildings, to achieve automatic inspection, maintenance, and repair. The swarms can disperse for wide-area inspection, and quickly reconfigure into a highly adaptive combination after detecting, cooperating to carry out repair operations.

(2) Disaster Rescue and Extreme Environment Detection
Reconfigurable swarms have obvious advantages in dealing with sudden disasters and exploring complex terrain environments. For example, in applications such as mine rescue and underground detection, reconfigurable swarms can delve into complex spaces, perform autonomous navigation, detect harmful gases, search for trapped personnel, etc. Their flexible swarm reconfiguration capability allows them to adapt to various narrow spaces, improving task success rates.

(3) Intelligent Manufacturing and Industrial Automation
Reconfigurable swarms can drive the upgrade of intelligent manufacturing in the era of Industry 4.0. For example, in smart warehousing and logistics, reconfigurable swarms can be widely used in intelligent logistics systems to optimize warehouse management and goods transportation. For instance, small units can independently transport small items, and when dealing with large goods or long-distance transportation, they can quickly reconfigure into a coordinated transportation modality, improving the intelligence and responsiveness of the logistics system.

(4) National Defense Security and Autonomous Operations
In the future of unmanned operations and national defense security, the flexibility and adaptability of reconfigurable swarms can be applied to battlefield reconnaissance and surveillance. Small reconfigurable swarms can cooperate in various modalities to infiltrate enemy areas for covert reconnaissance, and when needed, combine into a more powerful tactical platform to carry out autonomous investigation, information collection, and target identification tasks.

(5) Space Exploration and Extraterrestrial Operations
Reconfigurable swarms will also play an important role in space exploration and extraterrestrial operations. For example, in extraterrestrial exploration missions, reconfigurable swarms can adapt to different terrains in various modes, disperse to search for resources, or combine into a more stable transportation platform under special circumstances, improving the success rate of detection tasks.

1.3.2 Emerging Research Directions of Reconfigurable Swarms

Although reconfigurable swarms have shown great potential in multiple fields, they still need to combine emerging breakthrough technologies and a large amount of experimental research to promote their application in future real environments, such as:

(1) Utilize artificial intelligence and first-principle to construct multimodal mechanical models that are closer to complex physical dynamics, fully considering the impact of inter-swarm interactions and modal dynamics switching. Use advanced control methods to solve various adaptability problems that may exist in complex

environment applications, such as dense unstructured obstacles, sensor measurement errors, and unstable communications, etc. Explore in depth the improvements of models and optimization of control strategies under more complex, nonlinear environments.

(2) Develop docking mechanisms with motion degrees of freedom and multi-faceted reconfiguration capabilities, enabling reconfigurable swarms to have more diverse combination structures. This allows for more flexible adaptation to complex environments and diversified tasks, and conducts in-depth research on the control of individual units and combined modalities, reconfiguration docking strategies, and scheduling of multimodal behaviors, etc.

(3) Focusing on the high computational complexity and weak intelligent decision-making capabilities in large-scale swarm cooperation, research can be conducted on a distributed learning framework for multi-agent aimed at reconfigurable swarms. Surrounding the timeliness of large-scale swarm reconfiguration and the improvement of overall swarm cooperation efficiency, lightweight learning algorithms and intelligent optimization strategies are studied. Parallel and distributed computing are used to enhance the operability and real-time performance of the multimodal control framework, enabling it to maintain efficient autonomous cooperation in rapidly changing complex environments.

(4) Given that current applications are still not fully detached from the ideal laboratory environment and lack a deeper exploration of real scenarios and tasks, we are committed to applying reconfigurable swarms to more complex and realistic operational environments and tasks. For example, we are conducting robust algorithm verification and strategy optimization in extreme disturbance environments (such as sudden strong winds, drastic temperature changes, etc.), and swarm cooperation and configuration adaptation in rugged road conditions and harsh environments, etc., to further optimize the practical reliability of this swarm and develop an application framework for complex and changing environments. This not only helps to push the frontier exploration of theory and practice integration, but also provides important support for future commercial applications.

References

1. Quif1onez, Y., Baca, J., Delope, J., Ferre, M., Aracil, R.: Self-alignment approach based on cooperative behaviors for the docking process of modular mobile robots. In: Proc. 2010 IEEE Electron. Robot. Automot. Mech. Conf. (CERMA), pp. 445–450 (2010)
2. Wei, H., Chen, Y., Tan, J., Wang, T.: Sambot: A self-assembly modular robot system. IEEE/ASME Trans. Mechatron. **16**(4), 745–757 (2011)
3. Sun, X., Ge, W., Wang, X., Liu, J.: A reconfiguration approach for self-reconfigurable modular robot using assisted modules. In: Proc. 2015 IEEE Int. Conf. Mechatronics Autom. (ICMA), pp. 1436–1441 (2015)
4. Peck, R.H., Timmis, J., Tyrrell, A.M.: Self-assembly and self-repair during motion with modular robots. Sensors **23**(5), 2198 (2023)
5. Zhou, J., Wang, J., He, J., Gao, J., Yang, A., Hu, S., Li, Q.: Design, fabrication, and control algorithm of self-reconfigurable modular intelligent vehicles. Appl. Sci. **12**(14), 6886 (2022)

6. Cheah, W., Groves, K., Martin, H., Peel, H., Watson, S., Marjanovic, O., Lennox, B.: MIRRAX: A reconfigurable robot for limited access environments. IEEE Trans. Robot. **39**(2), 1341–1352 (2023)
7. Kutzer, M.D.M., Moses, M.S., Brown, C.Y., Armand, M., Scheidt, D.H., Chirikjian, G.S.: Design of a new independently-mobile reconfigurable modular robot. In: Proc. 2010 IEEE Int. Conf. Robot. Autom. (ICRA), pp. 2758–2764 (2010)
8. Wolfe, K.C., Moses, M.S., Kutzer, M.D.M., Chirikjian, G.S.: M3Express: A low-cost independently-mobile reconfigurable modular robot. In: Proc. 2012 IEEE Int. Conf. Robot. Autom. (ICRA), pp. 14–18 (2012)
9. Wang, W., Yu, W., Zhang, H.: JL-2: A mobile multi-robot system with docking and manipulating capabilities. Int. J. Adv. Robot. Syst. **7**(1), 9–18 (2010)
10. Wang, W., Lei, Z.: Trajectory planning for a reconfigurable mobile robot in mobile manipulating mode. In: Proc. 11th World Congr. Intell. Control Autom. (WCICA), pp. 42–47 (2014)
11. Zhang, X., Li, H., Xu, Y., Wang, X., Liu, H.: Self-reconfigurable modular robot based on graph theory and hybrid reconfiguration algorithms. Sensors **23**(78), 1–15 (2023)
12. Moubarak, P.M., Ben-Tzvi, P.: A tristate rigid reversible and non-back-drivable active docking mechanism for modular robotics. IEEE/ASME Trans. Mechatron. **19**(3), 840–851 (2014)
13. Sohal, S.S., Sebastian, B., Ben-Tzvi, P.: Autonomous docking of hybrid-wheeled modular robots with an integrated active genderless docking mechanism. J. Mech. Robot. **14**(1), 011010 (2022)
14. Park, C., Bae, J., Ryu, S., Lee, J., Seo, T.: R-Track: Separable modular climbing robot design for wall-to-wall transition. IEEE Robot. Autom. Lett. **6**(2), 1036–1042 (2021)
15. Leggieri, S., Canali, C., Caldwell, D.G.: Design of the crawler units: Toward the development of a novel hybrid platform for infrastructure inspection. Appl. Sci. **12**(11), 5579 (2022)
16. Liang, G., Luo, H., Li, M., Qian, H., Lam, T.L.: FreeBOT: A freeform modular self-reconfigurable robot with arbitrary connection point-design and implementation. In: Proc. 2020 IEEE/RSJ Int. Conf. Intell. Robots Syst. (IROS), pp. 6506–6513 (2020)
17. Zhao, D., Luo, H., Tu, Y., Meng, C., Lam, T.: Snail-inspired robotic swarms: a hybrid connector drives collective adaptation in unstructured outdoor environments. Nat. Commun. **15**, 3647 (2024)
18. Tu, Y., Liang, G., Lam, T.L.: FreeSN: A freeform strut-node structured modular self-reconfigurable robot-design and implementation. In: Proc. 2022 IEEE Int. Conf. Robot. Autom. (ICRA), pp. 7996–8002 (2022)
19. Lashkari, N., Biglarbegian, M., Yang, S.X.: Development of a novel robust control method for formation of heterogeneous multiple mobile robots with autonomous docking capability. IEEE Trans. Autom. Sci. Eng. **17**(4), 1759–1776 (2020)
20. Liu, C., Lin, Q., Kim, H., Yim, M.: SMORES-EP, a modular robot with parallel self-assembly. J. Mech. Robot. **13**(6), 4231–4238 (2023)
21. Min, H., Liu, Z., Wei, H., Wang, T.: Control of a self-assembly modular robot system over a wireless ZigBee network. In: Proc. 8th World Congr. Intell. Control Autom. (WCICA), pp. 1056–1062 (2010)
22. Wolfe, K.C., Kutzer, M.D.M., Armand, M., Chirikjian, G.S.: Trajectory generation and steering optimization for self-assembly of a modular robotic system. In: Proc. 2010 IEEE Int. Conf. Robot. Autom. (ICRA), pp. 4996–5001 (2010)
23. Sang, A.W.Y., Moo, C.G., Samarakoon, S.M.B.P., Muthugala, M.A.V.J., Elara, M.R.: Design of a reconfigurable wall disinfection robot. Sens. **21**, 6096 (2021)
24. Lin, X., Song, Y., Guo, L.: Short inter-robot distance following based on collision-free nonlinear model predictive control. In: Proc. 2024 6th Int. Conf. Reconfigurable Mech. Robot. (ReMAR), pp. 379–384 (2024)
25. Dorigo, M.: SWARM-BOT: An experiment in swarm robotics. In: Proc. 2005 IEEE Swarm Intell. Symp. (SIS), pp. 192–200 (2005)
26. Wei, R., Liu, Y., Dong, H., Zhu, Y., Zhao, J.: A graph-based hybrid reconfiguration deformation planning for modular robots. Sens. **23**(18), 7892 (2023)

27. Liang, G., Luo, H., Li, M., Qian, H., Lam, T.L.: FreeBOT: A freeform modular self-reconfigurable robot with arbitrary connection point-design and implementation. In: Proc. 2021 IEEE Int. Conf. Robot. Autom. (ICRA), pp. 1056–1062 (2021)
28. Ercan, H., Boyraz, P.: Design of a modular mobile multi-robot system: ULGEN (universal-generative robot). In: Proc. 2016 Asia-Pacific Conf. Intell. Robot Syst. (ACIRS), pp. 8–15 (2016)
29. Shi, Y., Elara, M.R., Le, A.V., Prabakaran, V., Wood, K.L.: Path tracking control of self-reconfigurable robot hTetro with four differential drive units. IEEE Robot. Autom. Lett. **5**(3), 3998–4005 (2020)
30. Li, J., Wu, H., Li, G., Chen, C., Xing, H.: Design and simulation of a self-reconfigurable modular wheel-legged robot. In: Proc. 2024 IEEE Int. Conf. Mechatronics Autom. (ICMA), pp. 145–158 (2024)
31. Zong, L., Liang, G., Lam, T.L.: Kinematics modeling and control of spherical rolling contact joint and manipulator. IEEE Trans. Robot. **39**(1), 738–752 (2023)
32. Fernando, A.H., Lim, L.A.: Symbiotic approach in a modular mobile robot. Int. J. Emerg. Trends Eng. Res. **9**(5), 602–607 (2021)
33. Leggieri, S., Canali, C., Caldwell, D.G.: Design, modeling, and experimental analysis of the Crawler Unit for inspection in constrained space. Annu. Rev. Control **57**, 100950 (2024)
34. Shi, Y., Elara, M.R., Kalimuthu, M., Devarassu, M.: sTetro: A modular reconfigurable cleaning robot. In: Proc. 2018 IEEE Int. Conf. Robot. Autom. (ICRA), pp. 6380–6386 (2018)
35. Sehgal, K., Upadhyaya, S., Keshav, Verma, M., Rayguru, M.M.: Backstepping based trajectory tracking control of a class of reconfigurable mobile robot. In: Proc. 2021 3rd Int. Conf. Adv. Comput. Commun. Control Netw. (ICAC3N), pp. 17–18 (2021)
36. Chwa, D.: Robust distance-based tracking control of wheeled mobile robots using vision sensors in the presence of kinematic disturbances. IEEE Trans. Ind. Electron. **63**(10), 6172–6183 (2016)
37. Ashrafiuon, H., Nersesov, S., Clayton, G.: Trajectory tracking control of planar underactuated vehicles. IEEE Trans. Autom. Control **62**(4), 1959–1965 (2017)
38. Sonker, R., Dutta, A.: Adding terrain height to improve model learning for path tracking on uneven terrain by a four-wheel robot. IEEE Robot. Autom. Lett. **6**(1), 239–246 (2021)
39. Li, H., Wang, H., Cui, L., Li, J., Wei, Q., Xia, J.: Design and experiments of a compact self-assembling mobile modular robot with joint actuation and onboard visual-based perception. Appl. Sci. **12**(6), 3050 (2022)
40. Pfotzer, L., Staehler, M., Hermann, A., Roennau, A., Dillmann, R.: KAIRO 3: Moving over stairs and unknown obstacles with reconfigurable snake-like robots. In: Proc. 2015 IEEE Int. Conf. Robot. Biomimetics (ROBIO), pp. 205–210 (2015)
41. Huang, B., Zhou, B., Pang, H., Xu, J., Fang, H.: A reliable docking mechanism and close-range docking algorithm for modular reconfigurable robots. In: Proc. 2023 9th Int. Conf. Mechatronics Robot. Eng. (ICMRE), pp. 78–83 (2023)
42. Tan, W., Wei, H., Yang, B.: SambotII: A new self-assembly modular robot platform based on Sambot. Appl. Sci. **8**(10), 1719 (2018)
43. Wei, H.X., Li, H.Y., Guan, Y., Li, Y.D.: A dynamics-based two-stage path model for the docking navigation of a self-assembly modular robot (Sambot). Robotica **34**(7), 1517–1528 (2016)
44. Jing, G., Tosun, T., Yim, M., Kress-Gazit, H.: An end-to-end system for accomplishing tasks with modular robots: perspectives for the AI community. In: Proc. 26th Int. Joint Conf. Artif. Intell. (IJCAI), pp. 4879–4883 (2016)
45. Chwa, D.: Fuzzy adaptive tracking control of wheeled mobile robots with state-dependent kinematic and dynamic disturbances. IEEE Trans. Fuzzy Syst. **20**(3), 587–593 (2012)
46. Chang, S., Wang, Y., Zuo, Z., Yang, H.: Fixed-time formation control for wheeled mobile robots with prescribed performance. IEEE Trans. Control Syst. Technol. **30**(2), 844–851 (2022)

Chapter 2
Reconfigurable Swarm System Design

Abstract The mechanical, hardware, and software design of reconfigurable swarms is introduced in this chapter, and the design of their basic unit motion, docking, separation, and combination motion functions is described. Moreover, the positioning, communication, and coordination platform required for the experiments in Chaps. 3–8 is established.

Keywords System construction · Functional verification · Platform development

2.1 Introduction

A reconfigurable swarm consists of several small mobile units with independent mobility and task execution capabilities [1–4]. Although the capabilities of individual units are limited, they can be reconfigured within the swarm to form various combined modalities suitable for different scenarios, achieving a balance between the capabilities and efficiency of the mobile units. In view of the practical needs of complex scenario operations, it is extremely necessary to study multi-modal motion control methods that are close to practical use and to build a practical reconfigurable swarm experimental system as a prerequisite. This can provide hardware support and model support for subsequent intra-modal control and inter-modal transition research.

2.2 Reconfigurable Swarm Experimental System Design

This book focuses on the study of the different modal motion characteristics and multi-modal motion control schemes of reconfigurable swarms in complex environments. This chapter constructs a reconfigurable swarm experimental system that includes the ontology structure, hardware system, and multi unit positioning and

T. Jiang et al., *Multi-modality Control of Reconfigurable Swarm*,
https://doi.org/10.1007/978-981-95-0305-6_2

communication. The basic motion capabilities of the swarm system and the feasibility of supporting subsequent multi-modal motion control algorithms are tested through various functional tests.

(1) Unit Ontology Mechanical Structure

The mechanical structure of the reconfigurable swarm unit is shown in Fig. 2.1, which includes the skeleton, shell, locomotion mechanisms and docking mechanisms. The overall design adopts a modular hardware structure, which is easy to assemble, replace, and transport. The modular unit skeleton uses lightweight, high-strength and tough carbon plates as the supporting material, ensuring the stability and safety of the core structure of the modular unit, and fixing threaded holes on the carbon plates to achieve the fixed connection of the hardware system, locomotion parts and shell. The modular unit shell is made of resin material and is quickly shaped using 3D printing technology, which not only ensures sufficient structural strength to protect internal elements, but also further ensures the lightweight of the overall structure. The locomotion mechanism consists of a spring suspension, a drive motor, and a wheel assembly. The spring suspension can ensure effective contact of the wheels on uneven roads, and the drive motor and its fixed 150 cm wheel together form the flexible mobility of the modular unit. The docking structure uses a mechanical locking method, composed of a positioning hole and a servo-controlled embedded cam lock, where the positioning hole is the passive docking end and the lock is the active docking end. When two modular units start docking, the lock of one unit enters the positioning hole of another unit, the stable combination of different modular units can be completed by rotating the lock through the servo, and at the same time, the active separation of the connected units can be achieved by rotating the lock through the servo.

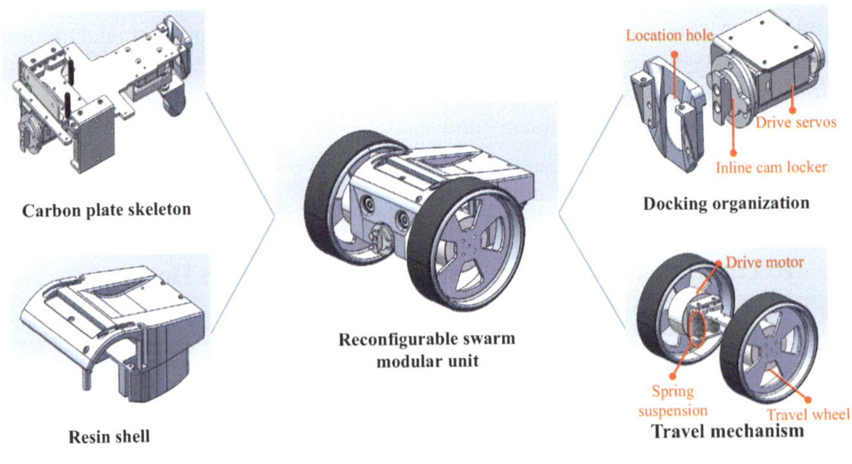

Fig. 2.1 Reconfigurable swarm mechanical structure

(2) Swarm unit Hardware System

The hardware system of the reconfigurable swarm unit is shown in Fig. 2.2, mainly composed of a control board, power system, communication module, power supply system, servo rudder module, and positioning marker. The control board is composed of the STM32F103ZET6 microcontroller and its peripheral circuits, responsible for receiving the desired velocity commands from the swarm host, sending the modular unit velocity status, generating the lower-level kinematic velocity control commands (desired torque current) and sending them to the torque motor. It is the core of the modular unit's lower-level control system. The STM32F103ZET6, as a low-cost, high-performance processor promoted by STMicroelectronics, has the advantages of small chip size, low system power consumption, and a rich set of general-purpose IO ports.

The control core board has a variety of interfaces, including 8.4 V, 5 V, and 3.3 V power interfaces (matching HQL010-B, R3008SB receiver and sensor power supply), CAN interface (matching MF7015 motor communication), USART interface (matching sensor communication and WiFi communication), SBUS interface (matching FUTABA-16SZ remote control communication), and variable voltage PWM interface (matching digital rudder communication).

The power system consists of two MF7015 torque motors controlled by a pot, which communicate with the modular unit control board through the embedded control module CAN bus, realizing real-time wheel velocity upload and control command (desired torque current) reception. The motor operates at a voltage of 24V, with an embedded zero-point permanent calibration 18-bit single-turn absolute value magnetic encoder, precision and low-temperature drift gold resistance sampling to ensure accurate current, low current and high resistance full N-tube three-phase full-bridge drive, real-time motor stator temperature detection circuit and high-performance 32-bit CORTEX M4 core ARM master control. The rated torque is $1\mathrm{N}\cdot\mathrm{m}$, the rated current is 8.5 A, and the torque, velocity, and position modes can be switched at will.

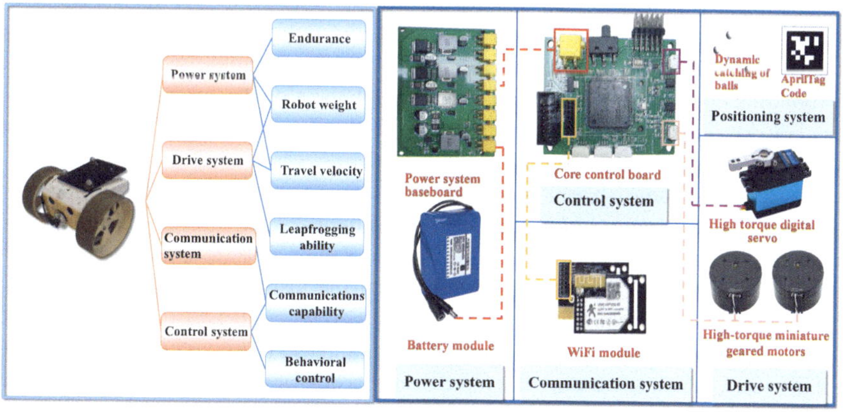

Fig. 2.2 Reconfigurable swarm unit hardware system

The desired torque current generated in the torque mode can be equivalent to the wheel velocity control torque of the modular unit in subsequent actual applications.

The communication module is a USR-WiFi1323-A2 manned WiFi module, with a 2.4 Ghz communication band, 3.3 V power supply, a maximum of 32 working modules, and a maximum transmission distance of 200 m under open view. It can use UART TTL serial communication to realize the data transmission and reception function between the upper-level control system and the lower-level control board.

The power supply system consists of a 24V lithium battery and a multi-channel buck power board. The lithium battery is responsible for the power supply of the entire system, and the power board converts the input 24V voltage into multiple standard voltages of different sizes to meet the power supply voltage requirements of different elements.

The servo rudder module is responsible for the position control of the auxiliary wheel and the docking mechanism rotary tightener, to realize the auxiliary wheel retraction and release and the docking mechanism lock and release before and after the modular unit's active reconfiguration and autonomous separation process. Its drive is PWM pulse mode, the working voltage is 5–6.5 V, and the controllable angle is 180°.

The positioning markers are of two types, Apriltag recognition and Marker reflective ball, which are applied to the subsequent two high-precision visual global positioning experiment schemes.

(3) Multi-unit Positioning/Communication/Swarm Control Host System

As a new type of reconfigurable multi-agent system, to achieve effective control of its clustering, docking, combination, and multi-modalities, it is necessary to design a host system that can obtain all modular unit information and schedule and control the swarm system. In this regard, this book provides two sets of solutions in the subsequent experimental process.

The first set, as shown in Fig. 2.3a, identifies the pose of swarm unit by a sensing system composed of a Hikvision high-precision MV-CA013-21UM grayscale camera and a 4 nm lens, which recognizes the Tag attached to its top. The captured image information can be quickly sent to the PC through the USB3.0 interface of the sensing system, and the precise positioning of the reconfigurable swarm unit and the surrounding environment can be obtained through the global pose solution algorithm based on AprilTag deployed by the host ROS. The pose information is published for subscription by the control node to be designed.

The reconfigurable swarm of this scheme can use the WiFi module, the host computer, and the router to form a local area network to realize the sharing of state information. The host computer uses the Lenovo Savior Y9000P-2021 model laptop, which can support the low-latency operation of the visual recognition positioning algorithm and the coordination algorithm. The wireless WIFI base station uses the Huawei AX3 router, and both the host computer and the WiFi module are connected to this router, which can ensure the rapid transmission of information of each unit of

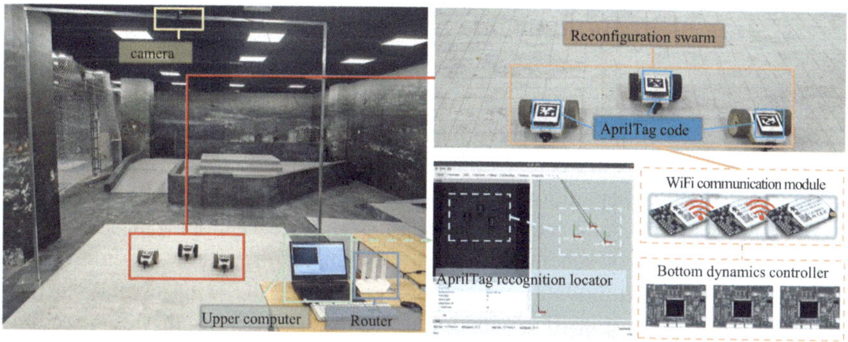

(a) High-precision global camera localization and communication scheme

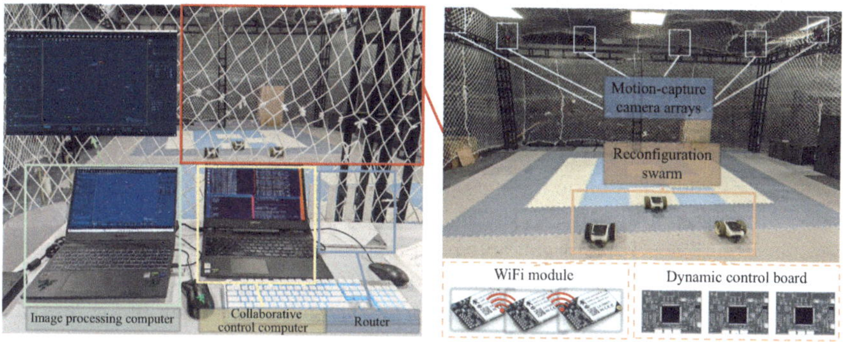

(b) Motion capture platform localization and communication scheme

Fig. 2.3 RS unit modality localization and swarm communication scheme

the reconfigurable swarm system. The gantry in the field is used to fix the camera, and the ground cloth is laid on the 3m × 3m field to reduce reflection and optimize image capture effects.

However, this scheme can only support the functional demonstration and effectiveness verification of the reconfigurable swarm in a small area due to the limited field of view of a single camera.

The second set is the NOKOV motion capture platform shown in Fig. 2.3b. During the experiment, the reflective ball Marker attached to the top of the modular unit is first recognized by the camera array, and then the NOKOV motion capture positioning algorithm is used to solve and obtain the pose of the modular unit. At this time, the control node of the host ROS system can subscribe to the modular unit pose published by the motion capture node of the image computer in the same local area network.

Combined with WiFi, the local area network where the host computer, coordination algorithm, and router are located, the velocity command generated by the host computer is sent to the bottom dynamics controller of the modular unit microcontroller to complete the design of the kinematics-dynamics cascaded closed-loop controller.

2.3 Testing of Reconfigurable Swarm Mobility Function

To verify the basic mobility and controllability of the reconfigurable swarm unit and the advantages of the combined modality's passability, this section sets up tests for single unit modality trajectory control, straight-line motion of dual unit combined modality, active separation of dual unit combined modality, and multi-terrain crossing ability of three-unit combined modality. These tests support the subsequent design of robust control schemes for reconfigurable unit swarm, reconfiguration docking, and combined modality progression, as well as the experimental verification of multi-modality transition scheduling. At the same time, the tests examine the swarm system's ability to traverse different terrains, highlighting its adaptability advantages in complex scene coordination operations.

Figure 2.4 is a scene diagram of the circular trajectory tracking of the reconfigurable swarm unit. The kinematics of the upper layer of the modular unit adopts the differential underdrive displacement motion unit backstepping-like control method proposed by Chaw [5], and the bottom dynamics adopts the traditional PID velocity control method. From the scene diagram and the trajectory tracking results, it can be seen that the system has basic motion control capabilities. In addition, other homogeneous driving control methods [6–9] are also available here. Figure 2.5 is a scene diagram of the linear motion and active separation test of the docking and combination of the two modular units of the reconfigurable swarm system. In the test, a linear motion velocity command is given. The system participating in the test can actively dock the unit to reach the fixed hole of the docking unit along a straight line, and judge the position status to actively reach the lock rotation command, complete the docking reconfiguration of the two modular units, and the combined modality after reconfiguration proceeds along a straight line trajectory, and timely reaches the separation command, the rear docking unit unlocks the locker, and the front unit actively moves forward in a straight line, completing the active separation of the two

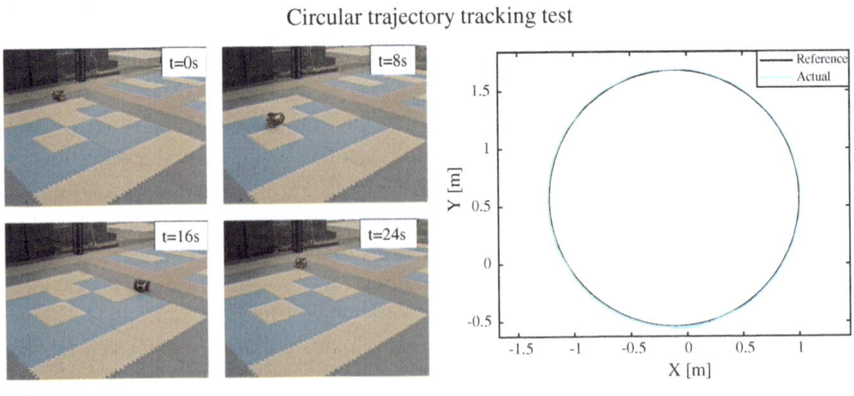

Fig. 2.4 Reconfigurable swarm unit modality trajectory control test

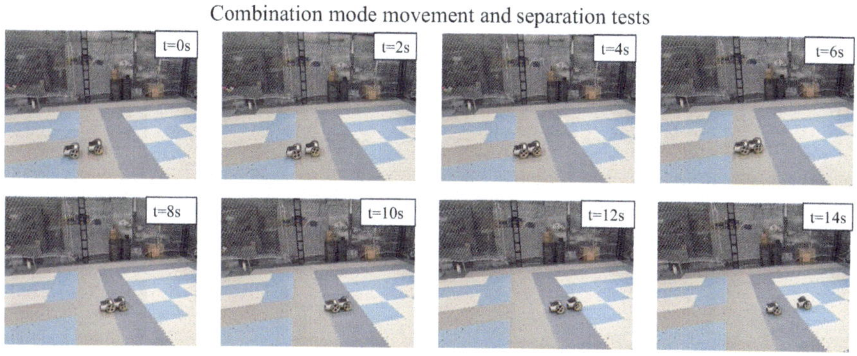

Fig. 2.5 Reconfigurable swarm unit modality docking, combination modality movement and separation tests

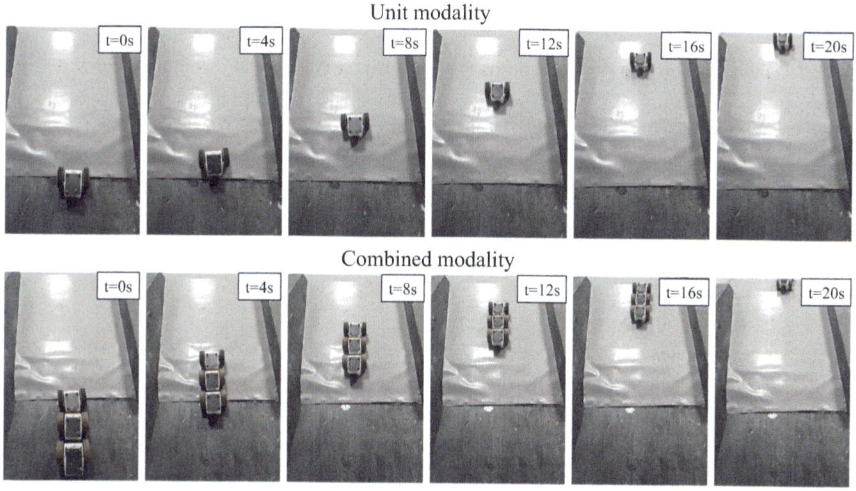

Fig. 2.6 Slope climbing test of the reconfigurable swarm unit and combination modalities

units. From this, it can be seen that the system has good combined modality motion performance, and the docking structure can flexibly realize the tasks of locking reconfiguration and release separation. The above test results of the modular unit and the combined modality can support the subsequent experimental verification of the modular unit formation, reconfiguration docking and combined modality advancement tasks and their modality switching and behavior scheduling in this book.

Figures 2.6, 2.7 and 2.8 respectively show the slope climbing, ditch crossing, and step climbing tests of the reconfigurable swarm unit and the combination of three units. Through these three typical terrain scenario tests, the passability performance of the unit and the combination can be obtained as shown in Table 2.1. The slope climbing ability of the reconfigurable swarm is verified by dynamically changing the

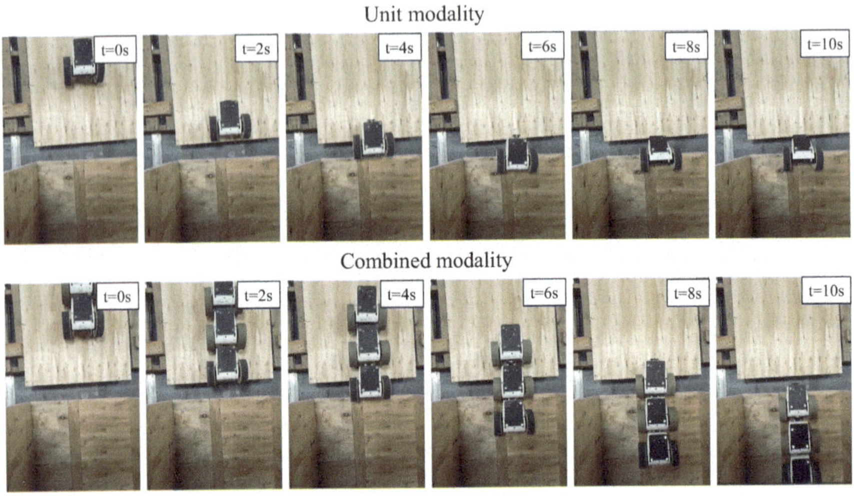

Fig. 2.7 Ditch crossing test of the reconfigurable swarm unit and combination modalities

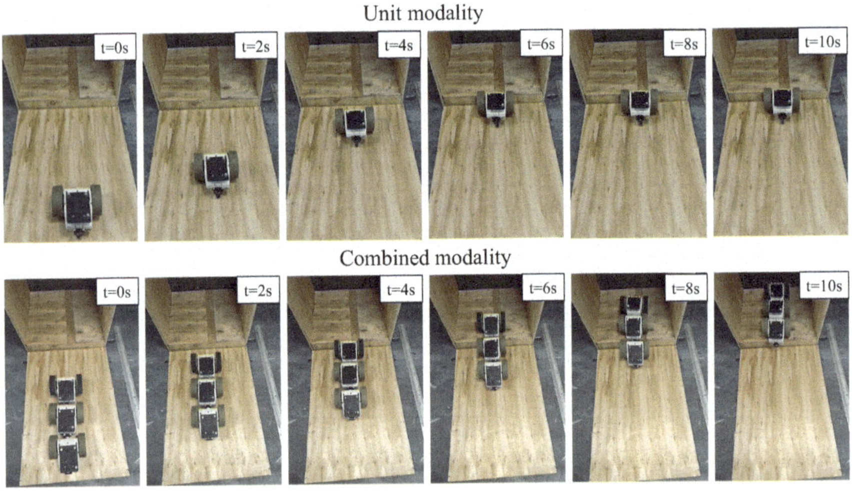

Fig. 2.8 Step climbing test of the reconfigurable swarm unit and combination modalities

1m slope to test the maximum slope climbing ability of the unit and the combination. The ditch crossing ability of the reconfigurable swarm is verified by dynamically changing the width of the ditch to test the maximum width of the ditch that the unit and the combination can pass through. The step climbing ability of the reconfigurable swarm is tested by dynamically changing the height of the step to test the maximum step height that the unit and the combination can cross. Through testing, it was found

Table 2.1 Terrain capability testing performance of unit and combination modality

Maximum passability	Unit	Combination
Maximum slope climbing (°)	24.24	34.98
Maximum ditch crossing width (cm)	8.0	12.0
Maximum step climbing height (cm)	3.2	10.2

that the combination has a significant improvement in slope climbing, ditch crossing, and step climbing abilities compared to the unit. It has a clear adaptive advantage in complex terrain scenarios, and compared to traditional fixed-configuration mobile vehicles, it has better application potential in complex dynamic scenarios. The reconfigurable swarm can in fact achieve bidirectional enhancement of capabilities and efficiency. Combined with the above docking, combination, and separation tests, it has the hardware basis for timely and demand-responsive convergence and divergence of capabilities.

2.4 Summary

This chapter introduces the construction of the reconfigurable swarm experimental system, including the introduction of the mechanical structure of the system unit, the hardware system, and the swarm communication scheme. It also carries out the capability and reliability verification of the modular unit and the combined modalities, supporting the subsequent theoretical research and method verification of modality internal control and modality transition.

References

1. Petersen, K.H., Napp, N., Stuart-Smith, R., Rus, D., Kovac, M.: A review of collective robotic construction. Sci. Robot. **4**(28), eaau8479 (2019)
2. Ozkan-Aydin, Y., Goldman, D.I.: Self-reconfigurable multilegged robot swarms collectively accomplish challenging terradynamic tasks. Sci. Robot. **6**(56), abf1628 (2021)
3. Liu, C., Lin, Q., Kim, H., Yim, M.: SMORES-EP, a modular robot with parallel self-assembly. Auton. Robots **47**(2), 211–228 (2023)
4. Paulos, J., Eckenstein, N., Tosun, T., Seo, J., Davey, J., Greco, J., Kumar, V., Yim, M.: Automated self-assembly of large maritime structures by a team of robotic boats. IEEE Trans. Autom. Sci. Eng. **12**(3), 958–968 (2015)
5. Chwa, D.: Tracking control of differential-drive wheeled mobile robots using a backstepping-like feedback linearization. IEEE Trans. Syst. Man Cybern. A **40**(6), 1285–1295 (2010)
6. Stefek, A., Pham, T.V., Krivanek, V., Pham, K.L.: Energy comparison of controllers used for a differential drive wheeled mobile robot. IEEE Access **8**, 170915–170927 (2020)

7. Chung, Y., Park, C., Harashima, F.: A position control differential drive wheeled mobile robot. IEEE Trans. Ind. Electron. **48**(4), 853–863 (2001)
8. Wang, J., Wang, Z., Liu, P., Xu, H., Guo, D., Wei, W.: Differential drive collaborative steering control of independent-wheel-drive articulated-steering electric vehicles for energy saving. IEEE Trans. Transp. Electrific. **10**(2), 4142–4158 (2024)
9. Štefek, A., Pham, V.T., Krivanek, V., Pham, K.L.: Optimization of fuzzy logic controller used for a differential drive wheeled mobile robot. Appl. Sci. **11**(13), 6023 (2021)

Chapter 3
Unit Modality Modeling and Disturbance Rejection Tracking Control

Abstract The kinematic-dynamic model of the reconfigurable swarm module unit is established in this chapter, and a robust kinematic-dynamic cascaded controller based on the hierarchical low-peak extented state observers is designed to achieve precise trajectory control of the system unit.

Keywords Modular unit · Mathematical modeling · Cascaded control · Disturbance rejection

3.1 Introduction

Reconfigurable swarm systems are composed of several mobile units, and their kinematic and dynamic models are the theoretical basis for developing unit control algorithms. It is essential to construct accurate kinematic and dynamic models of the units in a reconfigurable swarm system. In this chapter, we use a differential motion unit modality to describe the kinematic and dynamic characteristics of differential actuation, and establish a general kinematic and dynamic model for the units in a reconfigurable swarm system (similar to other homology-driven kinematic models [1, 2]). At the same time, considering the adverse impact of factors such as external disturbances and modeling uncertainties [3–5] that the reconfigurable swarm system may face in the actual environment and the inevitable peaking phenomenon in disturbance estimation [6–10], we design a disturbance rejection controller to cope with the disturbances and uncertainties in the external environment. Finally, using Lyapunov stability theory, we prove that the system has globally uniformly bounded exponential convergence.

T. Jiang et al., *Multi-modality Control of Reconfigurable Swarm*,
https://doi.org/10.1007/978-981-95-0305-6_3

3.2 Kinematic and Dynamic Model Construction for Reconfigurable Swarm Unit Modality

3.2.1 Unit Kinematic Model Construction

Unit kinematics focuses on studying the relationship between the position and geometric parameters of the unit in its workspace, as well as the constraint problems faced in trajectory planning. The derivation of the kinematic equations is influenced by the geometric structure of the unit, and its study provides the basis and premise for dynamics, stability analysis, and unit control. The modular units in the reconfigurable swarm system presented in this book adopt a differential actuation configuration, the geometric shape and motion parameters of which are shown in Fig. 3.1. The angular positions and angular velocities of the left and right wheels of each unit are θ_{il} and θ_{ir}, respectively. The linear velocities of the left and right wheels are defined as v_{il} and v_{ir}, respectively. The pose (position/direction) vector and its velocity in the inertial coordinate system $O_I X_I Y_I$ are $p_{ci} = [x_i, y_i, \theta_i]^{\mathrm{T}}$ (generalized coordinates of the modular unit) and $\dot{p}_{ci} = \left[\dot{x}_i, \dot{y}_i, \dot{\theta}_i\right]^{\mathrm{T}}$ (generalized velocities of the modular unit), respectively. The linear velocity and angular velocity of each modular unit are v_i and w_i, respectively. The direction of v_i is along the axis direction of the modular unit, coinciding with the X_v axis of the carrier coordinate system fixed on the modular unit. According to the right-hand rule, the Y_v axis of the carrier coordinate system can be determined. $\eta_i = [v_i, w_i]^{\mathrm{T}}$ is the composite velocity vector of the modular unit.

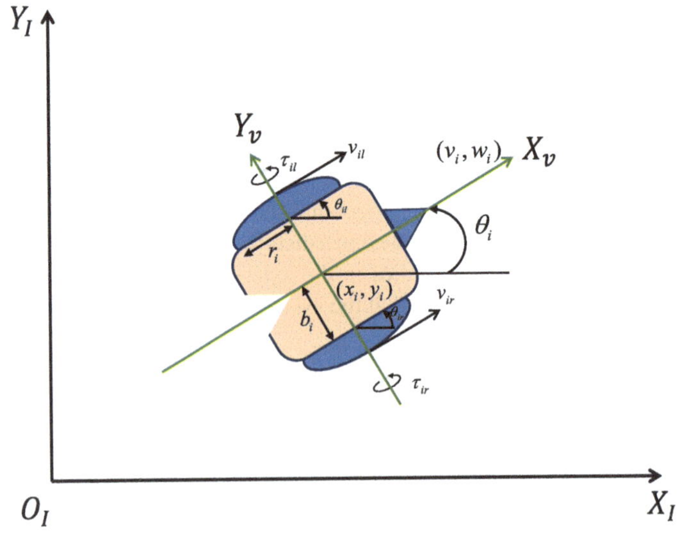

Fig. 3.1 RS unit modality coordinate diagram

Each reconfigurable swarm modular unit satisfies the following assumptions:

Assumption 3.1 The wheels on both sides of the reconfigurable swarm modular unit perform pure rolling without slipping.

Assumption 3.2 The steering axis is perpendicular to the plane $O_I X_I Y_I$.

According to the modality structure of the modular unit, its wheel velocity and linear velocity satisfy the following constraints

$$v_{ir} = v_i + b_i\dot{\theta}_i, \, v_{il} = v_i - b_i\dot{\theta}_i, \tag{3.1}$$

where b_i is the distance from the wheel to the axis of the modular unit.

The sum and difference of v_{ir} and v_{il} are easily obtained as

$$v_i = \frac{1}{2}\left(v_{ir} + v_{il}\right), 2b_i\dot{\theta}_i = v_{ir} - v_{il}, \tag{3.2}$$

where, according to Assumption 3.1, we have $v_{ir} = r_i\dot{\theta}_{ir}$ and $v_{il} = r_i\dot{\theta}_{il}$, r_i is the wheel radius of each modular unit, thus it is easy to derive

$$\begin{aligned} \dot{x}_i &= \frac{r_i}{2}\left(\theta_{ir}\cos\theta_i + \theta_{il}\cos\theta_i\right), \\ \dot{y}_i &= \frac{r_i}{2}\left(\theta_{ir}\sin\theta_i + \theta_{il}\sin\theta_i\right), \\ \dot{\theta}_i &= \frac{r_i}{2b_i}\left(\dot{\theta}_{ir} - \dot{\theta}_{il}\right), \end{aligned} \tag{3.3}$$

Combining Eq. (3.2), we can obtain the following kinematic model of the modular unit

$$\dot{p}_{ci} = S\left(p_{ci}\right)\eta_i, \tag{3.4}$$

where

$$S\left(p_{ci}\right) = \begin{bmatrix} \cos\theta_i & 0 \\ \sin\theta_i & 0 \\ 0 & 1 \end{bmatrix}, \eta_i = \begin{bmatrix} v_i \\ w_i \end{bmatrix}.$$

From Eq. (3.4), we can summarize the following nonholonomic constraint equation about the pose vector p_{ci}

$$A\left(p_{ci}\right)\dot{p}_{ci} = 0, \tag{3.5}$$

where $A\left(p_{ci}\right) = [\sin\theta_i, -\cos\theta_i, 0]$ is the nonholonomic constraint matrix, which will be used for the unconstrained transformation in the subsequent derivation of the dynamic model.

3.2.2 Modular Unit Dynamic Model Construction

After the kinematic modeling of the modular unit is completed, studying its dynamic modeling is the next key issue. Dynamic modeling is based on the physical requirements of inertia, elasticity, and friction. Its modeling can adopt various methods, but mainly according to the principles of Newton-Euler vector mechanics and Lagrange function method. The Lagrange function method replaces the cumbersome mechanical analysis through energy analysis, which is simpler and easier to understand than the Newton-Euler method, and is widely used in various dynamic model constructions. Therefore, this section uses the Lagrange method to construct the dynamic model of the reconfigurable swarm modular unit.

The centroid position of the modular unit is

$$r_{ci} = [x_i - a_i \cos\theta_i,\, y_i - a_i \sin\theta_i]^{\mathrm{T}} = [x_{ri},\, y_{ri}]^{\mathrm{T}},$$

the position derivative is $\dot{r}_{ci} = \frac{\partial r_{ci}}{\partial p_{ci}} \dot{p}_{ci} = T_i \dot{p}_{ci} = \left[\dot{x}_i + a_i\dot{\theta}_i \sin\theta_i,\, \dot{y} - a_i\dot{\theta}_i \cos\theta_i\right]^{\mathrm{T}}$, where a_i is the distance between the unit centroid and the wheelbase center, T_i is the corresponding Jacobian matrix

$$T_i = \begin{bmatrix} 1 & 0 & a_i \sin\theta_i \\ 0 & 1 & -a_i \cos\theta_i \end{bmatrix} \in \mathbf{R}^{2\times 3}.$$

The translational and rotational kinetic energy of the modular unit is

$$\begin{aligned} T\left(p_{ci}, \dot{p}_{ci}\right) &= \frac{1}{2}\left(m_i \dot{r}_{ci}^{\mathrm{T}} \dot{r}_{ci} + I_{mi}\dot{\theta}_i^2\right) \\ &= \frac{1}{2}\left(m_i \dot{p}_{ci}^{\mathrm{T}} T_i^{\mathrm{T}} T_i \dot{p}_{ci} + I_{mi}\dot{\theta}_i^2\right) \\ &= \frac{1}{2}\left(m_i\left(\dot{x}_i + a_i\dot{\theta}_i \sin\theta_i\right)^2 + m_i\left(y_i - a\dot{\phi}_i \cos\theta_i\right)^2 + I_{mi}\dot{\theta}_i^2\right) \\ &= \frac{1}{2}\dot{p}_{ci}^{\mathrm{T}} M_i\left(p_{ci}\right)\dot{p}_{ci}, \end{aligned} \tag{3.6}$$

where m_i and I_{mi} are the mass and moment of inertia of the modular unit, and $M_i\left(p_{ci}\right)$ is the inertia matrix. $M_i\left(p_{ci}\right) = m_i T_i^{\mathrm{T}} T_i + \mathrm{diag}\{0, 0, I_{m1}\}$, its specific matrix expression can be written as

$$M_i\left(p_{ci}\right) = \begin{bmatrix} m_i & 0 & a_i \sin\theta_i \\ 0 & m_i & -a_i \cos\theta_i \\ m_i a_i \sin\theta_i & -m_i a_i \cos\theta_i & m_i a_i^2 + I_{mi} \end{bmatrix}. \tag{3.7}$$

From the above equation, the derivative of the inertia matrix $M_i\left(p_{ci}\right)$ can be obtained as follows

$$\dot{M}_i\left(p_{ci}\right) = \begin{bmatrix} 0 & 0 & a_i\dot{\theta}_i\cos\theta_i \\ 0 & 0 & a_i\dot{\theta}_i\sin\theta_i \\ m_i\dot{\theta}_i a_i\cos\theta_i & m_i\dot{\theta}_i a_i\sin\theta_i & 0 \end{bmatrix}. \tag{3.8}$$

The Lagrange equation of the modular unit is

$$\frac{d}{dt}\left(\frac{\partial T_i}{\partial p_{ci}}\right) - \frac{\partial T_i}{\partial \dot{p}_{ci}} = Q_i + A^{\mathrm{T}}\left(p_{ci}\right)\lambda, \tag{3.9}$$

where Q_i is the generalized force vector relative to the generalized coordinates (orientation vector of the elemental body), λ is the multiplier of the Lagrange equation, $S^{\mathrm{T}}\left(p_{ci}\right)A^{\mathrm{T}}\left(p_{ci}\right) = [0, 0]^{\mathrm{T}}$.

From Eq. (3.6), we know that $\frac{\partial T_i}{\partial \dot{p}_{ci}} = M\dot{p}_{ci}$, then we have

$$\frac{d}{dt}\left(\frac{\partial T_i}{\partial \dot{p}_{ci}}\right) - \frac{\partial T_i}{\partial p_{ci}} = M\ddot{p}_{ci} + \dot{M}\dot{p}_{ci} - \frac{\partial T_i}{\partial \dot{p}_{ci}} = M\ddot{p}_{ci} + E\left(p_{ci}, \dot{p}_{ci}\right), \tag{3.10}$$

where $E\left(p_{ci}, \dot{p}_{ci}\right) = \dot{M}\dot{p}_{ci} - \frac{\partial T_i}{\partial p_{ci}}$ is the centrifugal and Coriolis terms.

From Eq. (3.10), the following $E_i\left(p_{ci}, \dot{p}_{ci}\right)$ can be calculated

$$E_i\left(p_{ci}, \dot{p}_{ci}\right) = \begin{bmatrix} m_i a_i^2\dot{\theta}_i^2\cos\theta_i \\ m_i a_i^2\dot{\theta}_i^2\sin\theta_i \\ e_{31} \end{bmatrix} = \begin{bmatrix} m_i a_i^2\dot{\theta}_i^2\cos\theta_i \\ m_i a_i^2\dot{\theta}_i^2\sin\theta_i \\ 0 \end{bmatrix}, \tag{3.11}$$

where the third term of E_i, $e_{31} = 0$. The unit vector of the module is $U_i = [\cos\theta_i, \sin\theta_i]^{\mathrm{T}}$. The left and right wheel actuation torques of each module are $\tau_i = [\tau_{ir}, \tau_{ir}]^{\mathrm{T}}$, and the wheel radius is r_i. The control force and torque vectors of each module are $F_i = \frac{\tau_{ir}+\tau_{il}}{r_i}U_i$ and $M_{ti} = \frac{\tau_{ir}-\tau_{il}}{r_i}b_i$. Define the position vector $r_{pi} = [x_i, y_i]$, the change in work done by the generalized force of the module is $\delta W_i = F_i^{\mathrm{T}}\delta r_{pi} + M_{ti}\theta_i$, where δr_{pi} and $\delta\theta_i$ are the rates of change of r_{pi} and θ_i. δW_i can be rewritten as

$$\delta W_i = F_i^{\mathrm{T}}\frac{\partial r_{pi}}{\partial p_{ci}}\delta p_{ci} + M_{ti}\frac{\partial\theta_i}{\partial p_{ci}}\delta p_{ci} = Q_i^{\mathrm{T}}\delta p_{ci}, \tag{3.12}$$

where $Q_i = \frac{\partial r_{pi}}{\partial p_{ci}}^{\mathrm{T}}F_i + \frac{\partial\theta_i}{\partial p_{ci}}^{\mathrm{T}}M_{ti}$, and the corresponding input injection matrix is $B_i\left(p_{ci}\right) = \frac{\partial Q_i}{\partial \tau_i}$, $\tau_i = [\tau_{ir}, \tau_{il}]^{\mathrm{T}}$.

According to Eq. (3.12), the specific forms of Q_i and $B_i\left(p_{ci}\right)$ of the module can be rewritten as

$$Q_i = \begin{bmatrix} \frac{\tau_{ir}+\tau_{il}}{r_i}\cos\theta_i \\ \frac{\tau_{ir}+\tau_{il}}{r_i}\sin\theta_i \\ \frac{\tau_{ir}-\tau_{il}}{r_i}b_i \end{bmatrix}, B_i\left(p_{ci}\right) = \begin{bmatrix} \frac{1}{r_i}\cos\theta_i & \frac{1}{r_i}\cos\theta_i \\ \frac{1}{r_i}\sin\theta_i & \frac{1}{r_i}\sin\theta_i \\ \frac{b_i}{r_i} & -\frac{b_i}{r_i} \end{bmatrix}. \tag{3.13}$$

Given $M_i\left(p_{ci}\right)$, $E_i\left(p_{ci}, \dot{p}_{ci}\right)$ and $B\left(p_{ci}\right)$, the constrained dynamics equation of the module can be obtained

$$M_i\left(p_{ci}\right)\ddot{p}_{ci} + E_i\left(p_{ci}, \dot{p}_{ci}\right) = B_i\left(p_{ci}\right)\tau_i + A^{\mathrm{T}}\left(p_{ci}\right)\lambda. \tag{3.14}$$

Combining the constraint condition of kinematics $S^{\mathrm{T}}\left(p_{ci}\right)A^{\mathrm{T}}\left(p_{ci}\right) = [0,0]^{\mathrm{T}}$, Eq. (3.14) can be transformed without constraint as follows

$$\bar{M}_{1i}\left(p_{ci}\right)\dot{\eta}_i + \bar{M}_{2i}\left(p_{ci}\right)\eta_i + \bar{E}_i\left(p_{ci}, \dot{p}_{ci}\right) = \bar{B}_i\left(p_{ci}\right)\tau_i, \tag{3.15}$$

where $\bar{M}_{1i}\left(p_{ci}\right) = S^{\mathrm{T}}\left(p_{ci}\right)M_i\left(p_{ci}\right)S\left(p_{ci}\right)$, $\bar{M}_{2i}\left(p_{ci}\right) = S^{\mathrm{T}}\left(p_{ci}\right)M_i\left(p_{ci}\right)\dot{S}\left(p_{ci}\right)$, $\bar{B}_i\left(p_{ci}\right) = S^{\mathrm{T}}\left(p_{ci}\right)B\left(p_{ci}\right)$ and $\bar{E}_i\left(p_{ci}, \dot{p}_{ci}\right) = S^{\mathrm{T}}\left(p_{ci}\right)E_i\left(p_{ci}, \dot{p}_{ci}\right)$.

The specific forms of the related matrices of the module dynamics equation can be written as

$$\bar{M}_{1i} = \begin{bmatrix} m_i & 0 \\ 0 & m_i a_i^2 + I_{mi} \end{bmatrix}, \bar{M}_{2i} = \begin{bmatrix} 0 & 0 \\ -m_i a_i^2 \dot{\theta}_i & 0 \end{bmatrix}, \bar{E}_i = \begin{bmatrix} m_i a_i^2 \dot{\theta}_i^2 \\ 0 \end{bmatrix}, \bar{B}_i = \begin{bmatrix} \frac{1}{r_i} & \frac{1}{r_i} \\ \frac{b_i}{r_i} & -\frac{b_i}{r_i} \end{bmatrix}. \tag{3.16}$$

3.3 Disturbance Rejection Cascade Control Under Uncertain Disturbance for Modular Unit

3.3.1 Introduction to Extended State Observer

Reconfigurable swarm systems often suffer from uncertainties and external disturbances during actual motion processes, and eliminating these adverse effects has become an important topic in controller design. Since most uncertain disturbances cannot be directly measured by sensors, the Extended State Observer (ESO) treats internal and external disturbances as lumped uncertainties and designs the overall disturbance as a new state variable. ESO uses the system's measured output variables or other measurable state variables, and observes the original internal states and extended states through the system input, without needing to construct a specific disturbance model or directly measure the disturbance itself, thus it is more adaptable and practical. In practical applications, if the overall disturbance and its rate of change are finite, there is no need to pay attention to its specific form, and it can be estimated in real time through ESO.

Consider the following system:

$$\dot{y} = f(y, \dot{y}, u, d) + b_0 u. \tag{3.17}$$

Its state space equation can be extended as:

$$\begin{cases} \dot{x}_1 = x_2, \\ \dot{x}_2 = x_3 + b_0 u, \\ \dot{x}_3 = h, \\ y = x_1, \end{cases} \tag{3.18}$$

where, the total disturbance can be extended state $x_3 = f, h = \dot{f}$ is the rate of change of the overall disturbance. Assuming h is bounded but finite, treating f as a state can estimate it in the form of a state observer, so for the above system, a nonlinear ESO can be designed as:

$$\begin{cases} \dot{\hat{x}}_1 = \hat{x}_2 + l_1(\hat{x}_1 - x_1), \\ \dot{\hat{x}}_2 = \hat{x}_3 + l_2 fal(\hat{x}_1 - x_1, \sigma_1, \delta_1) + b_0 u, \\ \dot{\hat{x}}_3 = l_3 fal(\hat{x}_1 - x_1, \sigma_2, \delta_2), \end{cases} \tag{3.19}$$

Here, $\hat{x}_3$ is the estimated value of the total disturbance, l_1, l_2, l_3 are the observer gains. $\sigma_1, \sigma_2, \delta_1$ and δ_2 are the parameters of the $fal(\cdot)$ function.

The $fal(\cdot)$ function is defined as follows:

$$fal(x_1 - \hat{x}_1, \sigma, \delta) = \begin{cases} \frac{x_1 - \hat{x}_1}{\delta - \sigma^2}, & |x_1 - \hat{x}_1| \le \delta, \\ |x_1 - \hat{x}_1|^{\sigma} \operatorname{sgn}(x_1 - \hat{x}_1), & |x_1 - \hat{x}_1| > \delta \end{cases} \tag{3.20}$$

Here, $\delta > 0$, and because the $fal(\cdot)$ function is a nonlinear function, it is generally referred to as a nonlinear ESO. The structure of the nonlinear ESO is complex, with many parameters that can only be adjusted by trial and error, making it difficult to implement. For ease of engineering implementation, it can be further designed through a linear ESO, the core idea of which is to design $\hat{x}_1 - x_1$ to replace the $fal(\cdot)$ function, and change the ESO to the traditional Luenberger observer form. Essentially, the linear ESO is a special case of the nonlinear ESO when $\sigma = 1$.

Rewrite Eq. (3.19) in the form of a state space equation:

$$\begin{cases} \dot{x} = Ax + Bu + Eh, \\ y = Cx, \end{cases} \tag{3.21}$$

where,

$$x = [x_1, x_2, x_3]^T, \quad A = \begin{bmatrix} 0 & 1 & 0 \\ 0 & 0 & 1 \\ 0 & 0 & 0 \end{bmatrix}, \quad B = \begin{bmatrix} 0 \\ 0 \\ 1 \end{bmatrix}, \quad E = \begin{bmatrix} 0 \\ 0 \\ 1 \end{bmatrix}, \quad C = [1, 0, 0].$$

The state space form of the linear ESO can be written as:

$$\begin{cases} \dot{\hat{x}} = A\hat{x} + Bu + L(y - \hat{y}), \\ \hat{y} = C\hat{x}, \end{cases} \tag{3.22}$$

Here, $L = [l_1, l_2, l_3]^T$ is the vector of observer gains.

The ESO used in this section is a linear ESO replacing a nonlinear ESO. The linear ESO is not only simple and easy to implement, but its control gain can also be directly selected by the root locus method, making the ESO gain obtain the characteristic roots of the characteristic polynomial of $A - LC$ equal to $-\omega_0$.

That is:

$$s^3 + l_1 s^2 + l_2 s + l_3 = (s + \omega_0)^3. \tag{3.23}$$

In this way, $A - LC$ becomes a Hurwitz polynomial. From this equation, we can get:

$$l_1 = 3\omega_0, \quad l_2 = 3\omega_0^2, \quad l_3 = \omega_0^3, \tag{3.24}$$

where, ω_0 is the only adjustable parameter of the ESO.

3.3.2 Low-Peak Extended State Observer Design Based on Kinematics

In practical applications, the reconfigurable swarm system may face a large amount of uncertainty and external disturbances when performing tasks in complex environments. Therefore, considering the disturbance situation, the kinematic equation of formula (3.4) can be changed to:

$$\begin{cases} \dot{x} = v\cos\theta + d_x, \\ \dot{y} = v\sin\theta + d_y, \\ \dot{\theta} = \omega, \end{cases} \tag{3.25}$$

where d_x, d_y are external disturbances.

Let the pose vector $r_p = \left[x_p, y_p, \theta_p\right]^T$, then:

$$\begin{cases} x_p = x - a\cos(\theta), \\ y_p = y - a\sin(\theta), \\ \theta_p = \theta, \end{cases} \tag{3.26}$$

where point P is the geometric center of the unit modality of the reconfigurable swarm system.

The derivative of the above is:

$$\begin{cases} \dot{x}_p = v\cos(\theta) + a\dot{\theta}\sin(\theta) + d_x, \\ \dot{y}_p = v\sin(\theta) - a\dot{\theta}\cos(\theta) + d_y, \\ \dot{\theta}_p = \dot{\theta}. \end{cases} \tag{3.27}$$

After rearranging, we get:

$$\dot{X} = \begin{bmatrix} \dot{x}_p \\ \dot{y}_p \end{bmatrix} = \begin{bmatrix} \cos(\theta) & a\sin(\theta) \\ \sin(\theta) & -a\cos(\theta) \end{bmatrix} \begin{bmatrix} v \\ w \end{bmatrix} + \begin{bmatrix} d_x \\ d_y \end{bmatrix}. \tag{3.28}$$

That is:

$$\dot{X} = U + D, \tag{3.29}$$

where

$$U = \begin{bmatrix} \cos(\theta) & a\sin(\theta) \\ \sin(\theta) & -a\cos(\theta) \end{bmatrix} \begin{bmatrix} v \\ w \end{bmatrix}, \quad D = \begin{bmatrix} d_x \\ d_y \end{bmatrix}.$$

From the above, we can get:

$$\begin{bmatrix} v_c \\ w_c \end{bmatrix} = \begin{bmatrix} \cos(\theta) & \sin(\theta) \\ -\sin(\theta) & \cos(\theta) \end{bmatrix} \begin{bmatrix} u_x \\ u_y \end{bmatrix}. \tag{3.30}$$

Let $X_1 = X$, $X_2 = D$, then:

$$\begin{cases} \dot{X}_1 = U + X_2, \\ \dot{X}_2 = H, \end{cases} \tag{3.31}$$

H is the derivative of D, $H = [h_x, h_y]^T$.

Therefore, the LPESO can be designed as follows:

$$\begin{cases} \dot{X}_1 = U + \hat{X}_2 + l_1(X_1 - \hat{X}_1), \\ \dot{X}_2 = l_2(X_1 - \hat{X}_1), \end{cases} \tag{3.32}$$

That is:

$$\dot{X} = A\hat{X} + BU + L\tilde{X}, \tag{3.33}$$

where:

$$A = \begin{bmatrix} 0 & 1 \\ 0 & 0 \end{bmatrix}, B = \begin{bmatrix} 1 \\ 0 \end{bmatrix}, L = \begin{bmatrix} l_1 \\ l_2 \end{bmatrix},$$

$$\hat{D} = \begin{cases} \hat{X}_2(t_k), & |e(t)| \geq \gamma(t), \\ \hat{X}_2(t_{k-1}), & |e(t)| < \gamma(t), \end{cases} \tag{3.34}$$

where $\hat{X} = [\hat{X}_1, \hat{X}_2]^T$ is the state variable of the LPESO, $\tilde{X} = X - \hat{X}$, $\hat{X}_2 = [\hat{d}_x, \hat{d}_y]^T$, $L = \text{diag}[l_1, l_2]$ is the observation gain of the observer, t_k is the time of time-triggering, $\hat{D}$ is the disturbance estimation output of the LPESO, $e(t) = C_0(X(t) - \hat{X}(t_k))$, C_0 is a constant vector, $\gamma(t) = ke^{-at} + b_0, k > 0, a > 0, b_0 > 0$. In general, $\gamma(t)$ is the first event trigger.

$$t_{k+1} = \inf\{t > t_k | e(t) \geq \gamma(t)\}, \tag{3.35}$$

where $\hat{X}_1, \hat{X}_2$ are estimates of X_1, X_2, l_1 and l_2 are observer gains, which can be selected by the pole placement method. The characteristic equation of the extended observer is:

$$(S + W_0)^2 = S^2 + 2W_0S + W_0^2, \tag{3.36}$$

where $S \in \mathbf{R}^{2\times1}$, $W_0 = [w_{x0}, w_{y0}]^T$ is the observer bandwidth, from which we can deduce:

$$\beta_1 = 2W_0, \quad \beta_2 = W_0^2. \tag{3.37}$$

Define the estimation error of the observer:

$$\begin{aligned} \tilde{X}_1 &= X_1 - \hat{X}_1, \\ \tilde{X}_2 &= X_2 - \hat{X}_2. \end{aligned} \tag{3.38}$$

The derivative of the estimation error of the observer is:

$$\begin{aligned} \dot{\tilde{X}}_1 &= \tilde{X}_2 - 2W_0\tilde{X}_1, \\ \dot{\tilde{X}}_2 &= h - W_0^2\tilde{X}_1. \end{aligned} \tag{3.39}$$

Let $\varepsilon = [\tilde{X}_1, \tilde{X}_2]^T$, the above observer error dynamics can be rewritten as:

$$\varepsilon = A\varepsilon + p(t), \tag{3.40}$$

where:

$$A = \begin{bmatrix} -2W_0 & 1 \\ -W_0^2 & 0 \end{bmatrix}, \quad p(t) = \begin{bmatrix} 0 \\ h(t) \end{bmatrix}.$$

It is easy to know that A is a Hurwitz matrix.

The effect of the kinematic low peak when given sinusoidal disturbance is shown in Fig. 3.2:

Lemma 3.1 *For Hurwitz matrix A, when $t > 0$, there exists a positive number c such that:*

$$\|e^{At}\| \leq ce^{\lambda_{\max}(A)t^2}$$

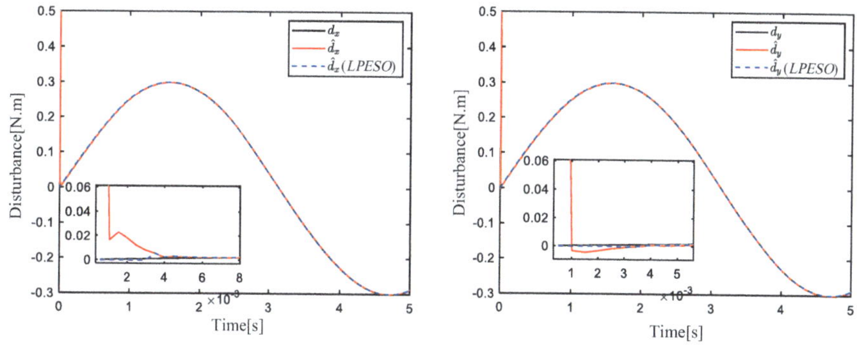

(a) Observation of Y-axis direction

(b) Observation of X-axis direction

Fig. 3.2 Observations of sinusoidal perturbations based on a kinematic inverse peak observer

Lemma 3.2 *Consider the subsystem (3.32), if $h(t)$ is bounded, then there exists a positive number σ, such that the observer estimation error:*

$$|\tilde{x}_e| \leq \sigma,$$

i.e., after a certain time, the model-based ESO is bounded.

3.3.3 Low-Peak Extended State Observer Design Based on Dynamics

In practical applications, considering the uncertainties and external disturbances of the mobile unit, the unit dynamics model in Eq. (3.15) is modified to:

$$\bar{M}_1(q)\dot{\eta} + \bar{M}_2(q)\eta + \bar{C}(q,\dot{q}) + F_f(\dot{q}) = \bar{B}(q)(\tau + \tau_d), \tag{3.41}$$

where $F_f(\dot{q}) \in \mathbf{R}^{2\times 1}$ is the nonlinear friction model of the modular unit, $\tau_d \in \mathbf{R}^{n\times 1}$ is the uncertain disturbance including unmodeled dynamics, $\tau \in \mathbf{R}^{n\times 1}$ is the torque output, $\bar{B}(q) \in \mathbf{R}^{n\times 1}$ is the input transformation matrix, $\lambda \in \mathbf{R}^{m\times 1}$ is the constraint force vector, $A(q) \in \mathbf{R}^{m\times n}$ is the constraint matrix, $\eta = [v, w]^T$.

For the system (3.41), introduce the control:

$$\bar{M}_1(q)u + \bar{M}_2(q)\eta + \bar{C}(q,\dot{q}) + F_f(\dot{q}) = \bar{B}(q)(\tau + \tau_d), \tag{3.42}$$

From Eqs. (3.41) and (3.42), eliminating the nonlinear terms, we can get:

$$\bar{M}_1(q)\dot{\eta} = \bar{M}_1(q)u. \tag{3.43}$$

Since $\bar{M}_1(q)$ is a symmetric positive definite matrix, $\bar{M}_1(q)$ is invertible, and we can obtain a decoupled constant system:

$$u = \dot{\eta}. \tag{3.44}$$

Introduce P control:

$$u = \dot{\eta}_c + K_p e, \tag{3.45}$$

where, $e = \eta_c - \eta$, $K_p > 0$ is the positive controller gain. Then the dynamics controller based on the model is as follows:

$$\tau = \bar{B}^{-1}\left[\bar{M}_1(q)(\dot{\eta}_c + K_p e) + \bar{M}_2(q)\eta + \bar{C}(q, \dot{q}) + F_f(\dot{q})\right] - \tau_d, \tag{3.46}$$

τ_d is the uncertainty part caused by modeling errors, parameter errors, etc., $\bar{\eta}_c = [v_c, w_c]^T$. In order to facilitate the design of the extended state observer, rewrite the velocity dynamics model:

$$\dot{\bar{\eta}} = \bar{M}_1^{-1}\left[\bar{B}(\tau + \tau_d) - (\bar{M}_2\bar{\eta} + \bar{C} + F_f)\right]. \tag{3.47}$$

That is:

$$\dot{\bar{\eta}} = \bar{M}_1^{-1}\bar{B}\tau + \bar{M}_1^{-1}\bar{B}\tau_d - \bar{M}_1^{-1}\left(\bar{M}_2\eta + \bar{C} + F_f\right). \tag{3.48}$$

Let $x_1 = \eta, x_2 = \bar{M}_1^{-1}\bar{B}\tau_d = d_f$, then the dynamics of the observed object are:

$$\dot{x}_1 = U + H + x_2, \tag{3.49}$$

where $U = \bar{M}_1^{-1}\bar{B}\tau, H = -\bar{M}_1^{-1}\bar{B}\left(\bar{M}_2\eta + \bar{C} + F_f\right)$, and the derivative of the extended state is $\dot{x}_2 = h(t) = \bar{M}_1\bar{B}\dot{\tau}_d$. In engineering, it can be assumed that the set is uncertain and its derivative is bounded. This chapter fully utilizes model information, weakens the impact of disturbances to a certain extent, and facilitates the extended state observer to achieve better results. According to the observed object, the following extended state observer is designed:

$$\begin{aligned} \dot{\hat{\bar{x}}}_1 &= U + H\bar{x}_2 + \beta_1(\bar{x}_1 - \hat{\bar{x}}_1), \\ \dot{\hat{\bar{x}}}_2 &= \beta_2(\bar{x}_1 - \hat{\bar{x}}_1), \end{aligned} \tag{3.50}$$

where $\hat{x} = [\hat{x}_1, \hat{x}_2]^T$ is the state variable of the low peak extended state observer, $\beta = [\beta_1, \beta_2]$ is the observation gain of the observer, t_k is the time of time-triggering, $\hat{D}$ is the disturbance estimation output of the LPESO. $e(t) = C_0(\hat{X}(t) - \hat{X}(t_k))$, C_0 is a constant vector. $\gamma(t) = ke^{-\alpha t} + b_0$, where $k > 0$, $\alpha > 0$, $b_0 > 0$. In general, t_0 is the first event trigger.

$$t_{k+1} = \inf\{t > t_k | e(t) \geq \gamma(t)\},$$

where $\hat{x}_1, \hat{x}_2$ are estimates of x_1, x_2, β_1 and β_2 are observer gains.

When a sinusoidal disturbance is given, the effect of the dynamic LPESO is shown in Fig. 3.3: To verify the effect of the LPESO in suppressing peaks in actual experiments, we deployed the LPESO in the bottom controller of the modular unit. The modular unit was suspended as shown in Fig. 3.4a, and the modular unit was moved according to the sinusoidal excitation trajectory. To more intuitively observe the harm of the peak, at a certain designated time during the operation of the modular unit, we designed a large wheel velocity error at this moment, and then used the ESO with a small observation gain to compare the traditional ESO and the LPESO designed in this chapter. The experimental results are shown in Fig. 3.4. It can be seen from the figure that the LPESO designed in this chapter has a good peak suppression effect.

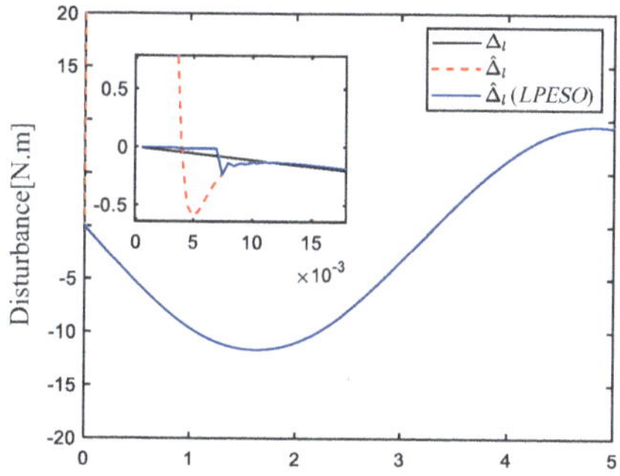

(a) Observation of left wheel

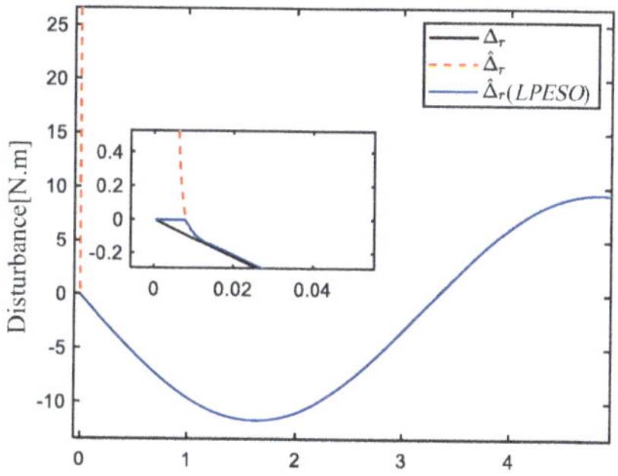

(b) Observation of right wheel

Fig. 3.3 Observations of sinusoidal perturbations based on a dynamic inverse peak observer

(a) Experimental scenario for comparison

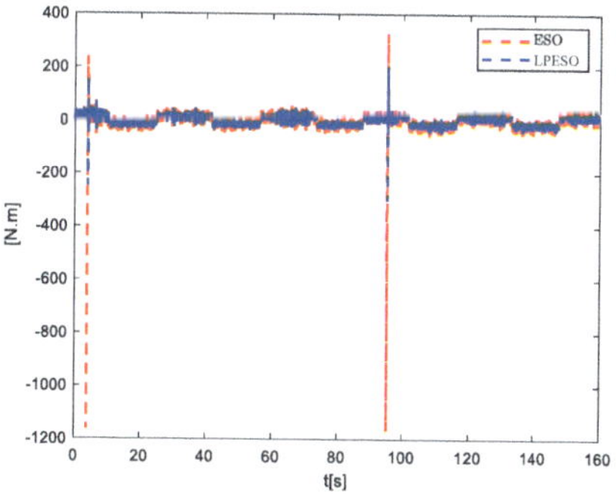

(b) Results of the comparative experiment

Fig. 3.4 Peak comparison experiment

3.4 Kinematic-Dynamic Cascaded Controller Based on Disturbance Feedforward Compensation

Combining Eqs. (3.27), (3.28), and (3.30), Eq. (3.25) can be rewritten as:

$$\dot{X} = \begin{bmatrix} \dot{x}_p \\ \dot{y}_p \end{bmatrix} = \begin{bmatrix} \cos(\theta) & a\sin(\theta) \\ \sin(\theta) & -a\cos(\theta) \end{bmatrix} \begin{bmatrix} v \\ w \end{bmatrix} + \begin{bmatrix} \hat{d}_x \\ \hat{d}_y \end{bmatrix}. \tag{3.51}$$

When the modular unit performs trajectory tracking control, the kinematic equation of the tracking trajectory is set as:

$$\dot{X}_d = \begin{bmatrix} \dot{x}_d \\ \dot{y}_d \end{bmatrix} = \begin{bmatrix} \cos(\theta_d) & a\sin(\theta_d) \\ \sin(\theta_d) & -a\cos(\theta_d) \end{bmatrix} \begin{bmatrix} v_d \\ w_d \end{bmatrix}. \tag{3.52}$$

where $[x_d, y_d, \theta_d]^T$ is the reference pose of the modular unit, v_d, w_d are the reference linear velocity and angular velocity of the modular unit.

Define new control variables u_v and u_θ:

$$\begin{bmatrix} v_c \\ w_c \end{bmatrix} = \begin{bmatrix} \cos(\theta) & a\sin(\theta) \\ \frac{1}{a}\sin(\theta) & -\frac{1}{a}\cos(\theta) \end{bmatrix} \begin{bmatrix} u_v - \hat{d}_x \\ w_\theta - \hat{d}_y \end{bmatrix}. \tag{3.53}$$

We can get:

$$\dot{x} = u_v, \quad \dot{y} = u_\theta.$$

The dynamic system in Eq. (3.42) is linear and decoupled, so the state feedback law is:

$$u_v = \dot{x}_d + c_1 x_e, \quad u_\theta = \dot{y}_d + c_2 y_e. \tag{3.54}$$

where, $x_e = x_d - x_p, y_e = y_d - y_p, \theta_e = \theta_d - \theta_p$. Combining Eqs. (3.53) and (3.54), we can obtain the overall nonlinear motion control rate as

$$\begin{bmatrix} v_c \\ w_c \end{bmatrix} = \begin{bmatrix} \cos(\theta) & \sin(\theta) \\ \frac{1}{a}\sin(\theta) & -\frac{1}{a}\cos(\theta) \end{bmatrix} \begin{bmatrix} \dot{x}_d + c_1 x_e - \hat{d}_x \\ \dot{y}_d + c_2 y_e - \hat{d}_y \end{bmatrix}. \tag{3.55}$$

The dynamics of the torque current control based on the LPESO can be designed as follows:

$$\tau = \tau_p - \tau_c, \tag{3.56}$$

$$\tau_p = \bar{B}^{-1}(\bar{M}_1(q)\dot{\eta}_c + K_p\eta_e) + \bar{M}_2(q)\eta + \bar{C}(q,\dot{q}) + F_r(\dot{q}), \tag{3.57}$$

$$\tau_c = \bar{M}_1^{-1}\bar{B}\hat{\tau}_d, \tag{3.58}$$

where τ_p is the torque current generated by the traditional calculation current torque method, and τ_c is the uncertainty compensation torque current generated by combining the LPESO.

3.5 Controller Stability Proof

Lemma 3.3 *For any given continuous time system's Lyapunov function $V(t)$, if $V(t)$ satisfies the following differential inequality:*

$$\dot{V} \leq -\lambda V + \varphi(t), \tag{3.59}$$

where λ is a positive constant, for $\forall t > 0$, when $\phi(t) = C$ is a constant or $\lim_{t\to\infty} \varphi(t) = C$, then the closed-loop system is globally ultimately bounded stable.

According to Eqs. (3.51) and (3.52), we have:

$$p_e = \begin{bmatrix} \dot{x}_d - \dot{x}_p \\ \dot{y}_d - \dot{y}_p \\ \dot{\theta}_d - \dot{\theta} \end{bmatrix} = \begin{bmatrix} \cos(\theta) & a\sin(\theta) \\ \sin(\theta) & -a\cos(\theta) \\ 0 & 1 \end{bmatrix} \begin{bmatrix} v_d - v \\ w_d - w \end{bmatrix} + \begin{bmatrix} \hat{d}_x \\ \hat{d}_y \\ 0 \end{bmatrix}. \tag{3.60}$$

Choose the Lyapunov function as

$$V_1 = \frac{1}{2}x_e^2 + \frac{1}{2}y_e^2. \tag{3.61}$$

Take the derivative of the Lyapunov function and substitute Eq. (3.60) into it, we get:

$$\begin{aligned} \dot{V}_1 &= x_e\dot{x}_e + y_e\dot{y}_e \\ &= x_e(\dot{x}_d - \dot{x}_p) + y_e(\dot{y}_d - \dot{y}_p) \\ &= x_e(-c_1x_e + \hat{d}_x - d_x) + y_e(-c_2y_e + \hat{d}_y - d_y) \\ &= -c_1x_e^2 - c_2y_e^2 + x_e(\hat{d}_x - d_x) + y_e(\hat{d}_y - d_y) \\ &= -c_1x_e^2 - c_2y_e^2 + x_e\tilde{x} + y_e\tilde{y}. \end{aligned} \tag{3.62}$$

From Lemma 3.2, we have

$$\begin{aligned} \dot{V}_1 &= -c_1x_e^2 - c_2y_e^2 + x_e\tilde{x} + y_e\tilde{y} \\ &\leq -c_1x_e^2 - c_2y_e^2 + x_e\sigma_x + y_e\sigma_y \\ &\leq -c_1x_e^2 - c_2y_e^2 + \left(\frac{x_e^2}{2} + \frac{\sigma_x^2}{2}\right) + \left(\frac{y_e^2}{2} + \frac{\sigma_y^2}{2}\right) \\ &= -\left(c_1 - \frac{1}{2}\right)x_e^2 - \left(c_2 - \frac{1}{2}\right)y_e^2 + \frac{\sigma_x^2 + \sigma_y^2}{2} \leq -\lambda V_1 + C_1, \end{aligned} \tag{3.63}$$

where $\lambda = \min\left\{c_1 - \frac{1}{2}, c_2 - \frac{1}{2}\right\}, c_1 > \frac{1}{2}, c_2 > \frac{1}{2}$, C_1 is a constant.

Let $V_2 = \frac{1}{2}\eta_e^T M_1 \eta_e$, then its derivative is:

$$\dot{V}_2 = \eta_e^T \bar{M}_1 \dot{\eta}_e = \eta_e^T \bar{M}_1 (\dot{\eta}_d - \dot{\eta}) = \eta_e^T \bar{M}_1 (\dot{\eta}_d - \dot{\eta}). \tag{3.64}$$

Then we have:

$$\begin{aligned}\dot{V}_2 &= \eta_e^T \bar{M}_1 \dot{\eta}_e \\ &= \eta_e^T M_1 (\dot{\eta}_c - \dot{\eta}) \\ &= \eta_e^T M_1 (\dot{\eta}_c - (M_1^{-1} B\tau + M_1^{-1} B\tau_d - M_1^{-1}(M_2\eta + C + F_r))).\end{aligned} \tag{3.65}$$

From the above equation, we can further obtain:

$$\begin{aligned}\dot{V}_2 &= \eta_e^T \bar{M}_1 (\dot{\eta}_c - (\dot{\eta}_c + k_p \eta_e + \bar{M}_1^{-1} \bar{M}_2(q)\eta + \bar{M}_1^{-1}\bar{C} + \bar{M}_1^{-1} F_f \\ &\quad - \bar{M}_1^{-1} B\dot{\hat{\tau}}_d + \bar{M}_1^{-1} B\tau_d - \bar{M}_1^{-1}(\bar{M}_2\eta + \bar{C} + F_f))) \\ &= \eta_e^T \bar{M}_1 (-k_p \eta_e - \bar{M}_1^{-1} B(\dot{\hat{\tau}}_d - \tau_d)) \\ &= -k_p \eta_e^T \bar{M}_1 \eta_e - \eta_e^T B(\dot{\hat{\tau}}_d - \tau_d) \le -k_p \eta_e^T \bar{M}_1 \eta_e - \eta_e^T B\sigma_0 \\ &\le -k_p \eta_e^T \bar{M}_1 \eta_e - \frac{\eta_e^T \eta_e}{2} + \frac{(\bar{B}\sigma_0)^T (\bar{B}\sigma_0)}{2} \le 0.\end{aligned} \tag{3.66}$$

Let $V = V_1 + V_2$, then:

$$V = \frac{1}{2}x_e^2 + \frac{1}{2}y_e^2 + \frac{1}{2}\eta_e^T \bar{M}_1 \eta_e \ge 0. \tag{3.67}$$

The derivative of the above equation is:

$$\begin{aligned}\dot{V} &\le -\left(c_1 - \frac{1}{2}\right)x_e^2 - \left(c_2 - \frac{1}{2}\right)y_e^2 + \frac{\sigma_x^2}{2} + \frac{\sigma_y^2}{2} - k_p \eta_e^T \bar{M}_1 \eta_e \\ &\quad - \frac{\eta_e^T \eta_e}{2} + \frac{(\bar{B}\sigma_0)^T \bar{B}\sigma_0}{2} = -\left(c_1 - \frac{1}{2}\right)x_e^2 - \left(c_2 - \frac{1}{2}\right)y_e^2 - \left(k_p + \frac{1}{2}\right)\eta_e^T \bar{M}_1 \eta_e \\ &\quad + \frac{(\bar{B}\sigma_0)^T \bar{B}\sigma_0}{2} + \frac{\sigma_x^2}{2} + \frac{\sigma_y^2}{2} \le -\lambda V + C_2,\end{aligned} \tag{3.68}$$

Here, λ is a constant, $\lambda = \min\left\{c_1 - \frac{1}{2}, c_2 - \frac{1}{2}, \frac{k_p}{2}\right\}$, C_2 is a constant, $\lim_{t\to\infty}\varphi(t) = C_2$, $\varphi(t) = -(B\sigma)^T B\sigma + \frac{\alpha x_e^2 + \beta y_e^2}{2}$. Here, $k_p = k_1 + k_2$, $k_v = k_1 + k_2$. According to Lemma 3.3, the closed-loop system is globally ultimately bounded stable, and the convergence rate of V is determined by λ.

3.6 Summary

This chapter focuses on the modeling and disturbance control of the modular unit of the reconfigurable swarm system. Firstly, based on the physical characteristics of the mobile unit driven by differential velocity, the kinematic and dynamic models of the modular unit of the reconfigurable swarm system are established. Through the analysis of non-integral constraints, the Lagrange method is used to derive the dynamic equations, and a simplified model is obtained through unconstrained transformation. Secondly, in response to the uncertainties and modeling errors in the actual environment, a cascaded disturbance control framework based on the low-peak Extended State Observer is proposed. At the kinematic level, the low-peak Extended State Observer is designed to estimate and compensate for external disturbances in real time; at the dynamic level, a disturbance observer is designed in combination modalities with model information to effectively suppress the initial peak phenomenon of the traditional ESO, and its superiority is verified through experiments. Finally, using Lyapunov theory, the global uniform ultimate boundedness of the closed-loop system is proven, and the effectiveness of the proposed method is verified through comparative experiments, providing reliable protection for the practical application of the reconfigurable swarm system.

References

1. Zhang, M., Chen, N., Wang, H., Qiu, J., Han, Z., Ren, Q.: Universal trajectory optimization framework for differential drive robot class. IEEE Trans. Autom. Sci. Eng. **22**, 13030–13045 (2025)
2. Allagui, N.Y., Salem, F.A., Aljuaid, A.M.: Artificial fuzzy-PID gain scheduling algorithm design for motion control in differential drive mobile robotic platforms. Comput. Intell. Neurosci. **2021**, 5542888 (2021)
3. Xie, H., Zheng, J., Chai, R., Nguyen, H.T.: Robust tracking control of a differential drive wheeled mobile robot using fast nonsingular terminal sliding mode. Comput. Electr. Eng. **96**, 107488 (2021)
4. Hassan, N., Saleem, A.: Neural network-based adaptive controller for trajectory tracking of wheeled mobile robots. IEEE Access **9**, 126497–126512 (2021)
5. Qin, B., Yan, H., Zhang, H., Wang, Y., Yang, S.X.: Enhanced reduced-order extended state observer for motion control of differential driven mobile robot. IEEE Trans. Cybern. **53**(2), 1299–1310 (2023)
6. Astolfi, D., Marconi, L., Praly, L., Teel, A.R.: Low-power peaking-free high-gain observers. Automatica **98**, 169–179 (2018)
7. Chitour, Y.: Time-varying high-gain observers for numerical differentiation. IEEE Trans. Autom. Control **47**(9), 1565–1569 (2002)
8. Lou, D.-K., Zhang, D., Liang, H.-G.: An improved high-gain observer design for flexible needle steering. IEEE Trans. Circuits Syst. II, Exp. Briefs **70**(9), 3489–3493 (2023)
9. Behera, A.K., Shim, H.: Robust feedback stabilization using high-gain observer via event triggering. Int. J. Robust Nonlinear Control **30**(15), 5816–5833 (2020)
10. Wang, H., Zuo, Z., Wang, Y., Yang, H.: Composite nonlinear path-following control for unmanned ground vehicles with anti-windup ESO. IEEE Trans. Syst. Man Cybern. Syst. **52**(9), 5865–5876 (2022)

Chapter 4
Unit Modality Formation Control Under Nested Motion Saturation

Abstract A hierarchical kinematic–dynamic anti-saturation formation control method is proposed in this chapter. At the kinematic level, a distributed control law based on nonholonomic constraints suppresses velocity and acceleration saturation. At the dynamic level, an adaptive saturation extended state observer compensates for lumped uncertainties. Simulations and experiments validate the effectiveness.

Keywords Unit modality · Nested motion saturation · Adaptive saturation suppression · Peaking rejection

4.1 Introduction

Reconfigurable swarm are special multi-mobile unit systems composed of subunits with independent mobility and execution capabilities. The subunits in the swarm can not only enhance their capabilities through recombination, but also disperse their formations to improve execution efficiency and operational range. There are few studies on reconfigurable swarm subunit formations, and although advanced control methods for homogeneously driven mobile units can be referred to, the operational environment is complex and harsh. In practical applications, it is necessary to fully consider various restrictive constraints in coordination and the negative impact of lumped disturbances. This chapter proposes a robust control framework for kinematics-dynamics cascaded formation of multi-unit systems under distributed network topology, considering the nonholonomic constraints, nested saturation of velocity and acceleration, and the lumped uncertainty disturbances. Firstly, through the cascaded control of position and heading in the kinematic layer, the leader-following and formation maintenance of the agents are realized. Compared with the existing nonholonomic formation control methods [1–3], the nested saturation constraints of velocity and acceleration are considered for the first time. An adaptive auxiliary dynamic system is designed based on the nonholonomic constraints and nested motion saturation, which is used to smooth the possible saturation oscillations of the formation trajectory. Secondly, an Adaptive Saturation Extended State

T. Jiang et al., *Multi-modality Control of Reconfigurable Swarm*,
https://doi.org/10.1007/978-981-95-0305-6_4

Observer (ASESO) is designed in the dynamic bottom layer. Compared with the traditional Extended State Observer (ESO) [4, 5], which is prone to peak phenomenon, it can achieve low-peak disturbance estimation and enhance the robustness of the cascaded dynamic velocity control through the feedforward compensation structure. Finally, through the comparative simulation and physical experiments of the agent formation, the proposed method is verified to enhance the adaptability in terms of smoothness and disturbance resistance.

4.2 Problem Description

Each unit of the reconfigurable swarm has autonomous operating capabilities, and the operating efficiency can be doubled through swarm coordination. The modular unit assembly is set to $m+1$, and the leader-follower coordination structure is adopted. The data communication topology of each follower unit is described by an undirected graph $\mathcal{G}=\{\mathcal{V},\mathcal{E}\}$, where the vertex set is $\mathcal{V}=\{r_i, i=1,\ldots,m\}$, and the edge set is $\mathcal{E}=\mathcal{V}\times\mathcal{V}$. The edge (r_j, r_i) represents the information flow from vertex j to vertex i. The weighted adjacency matrix is $\mathcal{A}=\left[a_{ij}\right]_{m\times m}$, where $a_{ii}=0$, if $\left(r_j, r_i\right)\in\mathcal{V}$ and $i\neq j$, then $a_{ij}=1$. The Laplacian matrix of the undirected graph is defined as $\mathcal{L}=\left[l_{ij}\right]_{m\times m}$, where $l_{ii}=\sum_{j\neq i}a_{ij}$ and $l_{ij}=-a_{ij}$. $\mathcal{B}=\operatorname{diag}\{b_1,\ldots,b_m\}\in\mathbf{R}^{m\times m}$ represents the communication weight between the ith follower and the leader r_0; if the ith follower can obtain the leader's information, then $b_i=1$, otherwise $b_i=0$. The extended graph including the leader and followers is defined as $\bar{\mathcal{G}}=\{\bar{\mathcal{V}},\bar{\mathcal{E}}\}$, where $\bar{\mathcal{V}}=\{r_0, r_1,\ldots,r_m\}$, $\bar{\mathcal{E}}=\bar{\mathcal{V}}\times\bar{\mathcal{V}}$.

The kinematic model of the modular unit is given by Eq. (3.4). According to Eqs. (3.1)–(3.3), the linear velocity and angular velocity of the ith swarm unit and its left-right wheel velocity conversion relationship are as follows [6]

$$\omega_{il}=\left(v_i-b_i w_i\right)/r_i, \omega_{ir}=\left(v_i+b_i w_i\right)/r_i, \tag{4.1}$$

where ω_{0k} $(k=l,r)$ is the velocity of the left and right wheels of the leader unit, and ω_{ik} $(i=1,\ldots,m; k=l,r)$ is the velocity of the left and right wheels of the follower units. The dual wheels of the modular unit are driven by a torque motor. Since the motor has embedded current loop control, the issued torque current command can be regarded as a dynamic control input proportional to the actual torque. In this chapter, combined with the dynamic model (3.16), velocity conversion (4.1) and the characteristics of the actuation wheel mechanism, the following low-level wheel velocity dynamic model of the modular unit is considered:

$$\begin{aligned}\tau_{ir}&=D_{11}\dot{\omega}_{ir}+D_{12}\dot{\omega}_{ir}+f_c\left(\omega_{ir}\right)+d_{ir},\\ \tau_{il}&=D_{21}\dot{\omega}_{il}+D_{22}\dot{\omega}_{il}+f_c\left(\omega_{il}\right)+d_{il},\end{aligned} \tag{4.2}$$

In the formula, $D_{11} = \frac{m_i r_i^2}{4} + \frac{I_{mi}}{4b_i} + I_{mli}, D_{12} = D_{21} = \frac{m_i r_i^2}{4} - \frac{I_{mi}}{4b_i}$, and $D_{22} = \frac{m_i r_i^2}{4} + \frac{I_{mi}}{4b_i} + I_{mri}$ are the inertia parameters of the reconfiguration swarm unit; d_{0l} and d_{0r} represent the total disturbance of the navigator, including the model uncertainty and external disturbance in the modular unit velocity dynamics; d_{il} and d_{ir}, $i = 1, \ldots, m$, represent the total disturbance of the ith follower in the modular unit velocity dynamics. Because the modular unit is designed to keep the center of gravity on the wheelbase as much as possible, it can be assumed that $a_i = 0$ in the bottom control model (4.2), even if there is a deviation between the center of gravity and the wheelbase center in the actual modality, it can also be treated as part of the model uncertainty of the total disturbance. In addition, the effect of driving wheel friction can be further introduced into the model, where the composite nonlinear model [7] realistically reflects the effects of static friction, viscous friction, Coulomb friction, and Stribeck friction in engineering applications. Therefore, the composite nonlinear friction effects $f(\omega_{il})$ and $f(\omega_{ir})$ are introduced to more comprehensively describe the modular unit bottom wheel velocity dynamics, as shown below

$$\begin{aligned} f_c(\omega_{ik}) = {} & \mu_{1k}\left(\tanh\left(\mu_{2k}\omega_{ik}\right) - \tanh\left(\mu_{3k}\omega_{ik}\right)\right) + \mu_{4k}\omega_{ik} \\ & + \mu_{5k}\tanh\left(\mu_{6k}\omega_{ik}\right), \end{aligned} \tag{4.3}$$

where $\mu_{1k}, \mu_{2k}, \mu_{3k}, \mu_{4k}, \mu_{5k}$, and μ_{6k} are the coefficients of static friction, Coulomb friction, viscous friction, and Stribeck friction effects, respectively. In this chapter, Eqs. (3.4) and (4.2) are used as the control models for subsequent coordinated formation and reconfiguration docking of the modular unit.

Referencing the amplitude and velocity nested saturation form [8], the velocity and acceleration nested saturation model is shown as follows

$$\begin{aligned} \dot{v}_{is} &= \mathrm{sat}_{\bar{a}}\left(\left(\mathrm{sat}_{\bar{v}}\left(v_{id}\right) - v_{is}\right)/\tau_v\right), \\ \dot{w}_{is} &= \mathrm{sat}_{\bar{\omega}}\left(\left(\mathrm{sat}_{\bar{w}}\left(w_{id}\right) - w_{is}\right)/\tau_w\right), \end{aligned} \tag{4.4}$$

and

$$\mathrm{sat}_\sigma(x) = \begin{cases} x, & |x| < \sigma \\ \sigma\,\mathrm{sgn}(x), & |x| \geq \sigma \end{cases},$$

where v_{0d} and w_{0d} are the virtual linear velocity and angular velocity of the leader; v_{id} and $w_{id}, i = 1, \ldots, m$, represent the virtual linear velocity and angular velocity of the ith follower, which are used as the input for subsequent kinematic control design. v_{is} and w_{is} are the virtual linear velocity and angular velocity under nested saturation; $\tau_v, \tau_w \in \mathbf{R}_+$ are the system time constants; $\bar{v}$ and $\bar{w}$ are the saturation thresholds of the linear velocity and angular velocity for each unit; $\bar{a}, \bar{\omega} \in \mathbf{R}_+$ are the saturation thresholds for the corresponding linear acceleration and angular acceleration. As known from Ref. [8], the amplitude and rate of change of the linear velocity and angular velocity commands of the swarm units can be obtained from the above nested saturation model.

The orthogonal coordinates (p_{ix}, p_{iy}), $i = 1, \ldots, m$, are used as the positions relative to each follower with respect to the leader, which are used to describe the formation configuration in multi-unit coordination. v_r and w_r represent the desired linear velocity and angular velocity of the leader, guiding the movement of the followers, and the corresponding reference trajectory can be expressed as follows

$$\begin{aligned} \dot{x}_r &= v_r \cos(\theta_r), \\ \dot{y}_r &= v_r \sin(\theta_r), \\ \dot{\theta}_r &= w_r, \end{aligned} \tag{4.5}$$

where x_r, y_r and θ_r are the position and heading of the reference trajectory.

Assumption 4.1 The total disturbances d_{il} and d_{ir} and their derivatives of each formation unit are bounded [9].

Assumption 4.2 The linear velocity and angular velocity of the leader's reference trajectory are consistent with the velocity and acceleration saturation model, and the reference configuration p_{ix} and p_{iy} and their derivatives remain bounded.

Assumption 4.3 The communication graph $\bar{\mathcal{G}}$ includes a directed spanning tree rooted at the leader r_0, and the matrix $\mathcal{M} = \mathcal{L} + \mathcal{B}$ is symmetric positive definite [10].

4.3 Smooth Formation Control Under Nested Velocity/Acceleration Saturation

This chapter adopts the multi-unit distributed formation control framework based on the leader-follower structure as shown in Fig. 4.1. Both the leader and the followers adopt a kinematic-dynamic cascaded control structure. In the kinematic layer, a position and heading cascaded structure is adopted, the longitudinal and lateral virtual velocities of the leader and the cascaded virtual angular velocity are designed to stabilize the position and heading tracking errors of the leader. The followers first design a distributed estimator based on the distributed communication topology to generate longitudinal and lateral position coordination errors and the position status of the leader; then, the followers' longitudinal and lateral virtual velocities are designed to stabilize the coordinated position errors and generate corresponding virtual linear velocities and headings; finally, the virtual angular velocity is designed to stabilize the heading tracking error. In the dynamic control layer, feedback controllers for the leader and followers based on the dynamic model are designed to track the wheel velocity commands converted from the virtual linear velocity and angular velocity, and integrate the adaptive auxiliary dynamic system and the adaptive saturation extended state observer to smooth tracking and formation trajectory oscillation and suppress unknown random disturbances, thus achieving robust and smooth formation of the multi-unit system with a kinematic-dynamic cascaded structure.

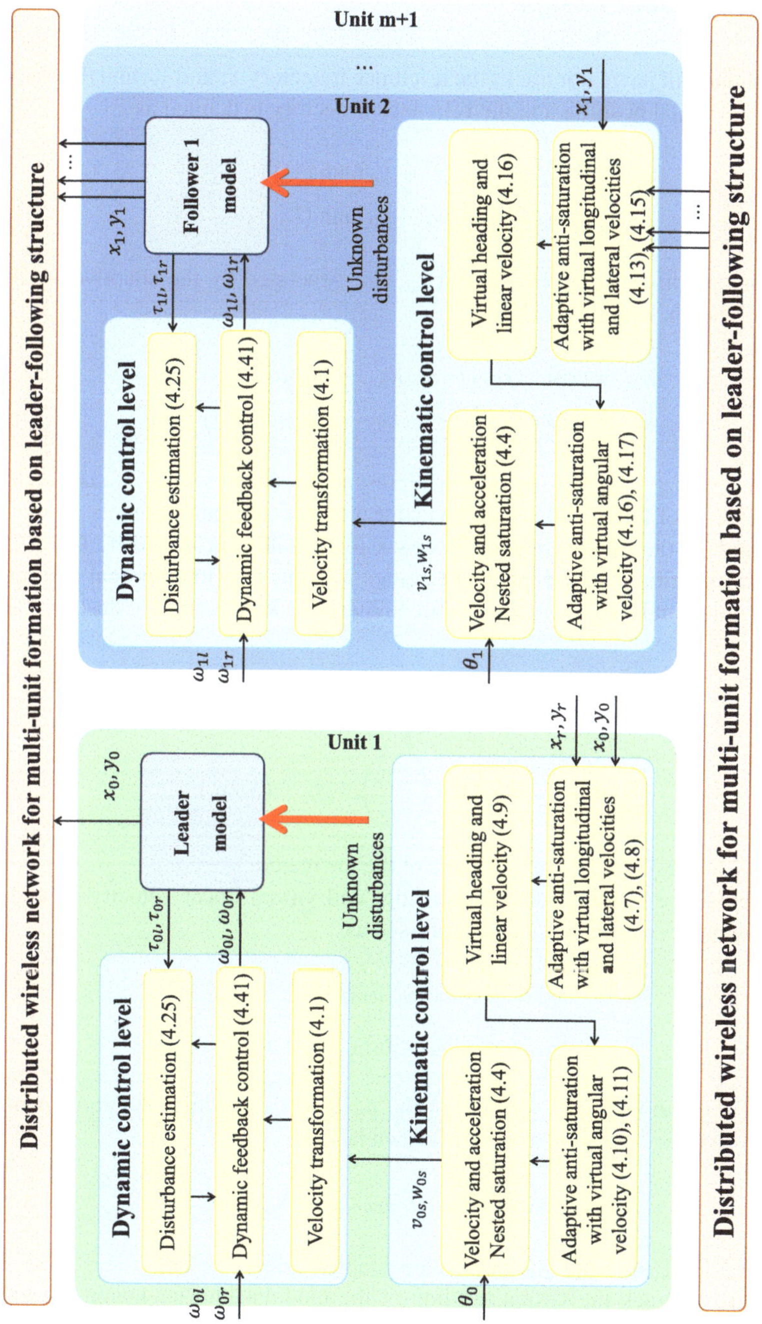

Fig. 4.1 Multi-agent coordination formation control framework

4.3.1 Adaptive Anti-saturation Distributed Formation Controller Design

The modular unit navigator tracks the reference trajectory x_r and y_r, and the tracking error of the actual position and the reference trajectory is defined as

$$\begin{aligned} e_{10} &= x_0 - x_r - \gamma_0 \tanh\left(\lambda_{0x}\right), \\ e_{20} &= y_0 - y_r - \gamma_0 \tanh\left(\lambda_{0y}\right), \end{aligned} \tag{4.6}$$

where the auxiliary variables λ_{0x} and λ_{0y} are generated by the adaptive auxiliary dynamic system, as shown below

$$\begin{aligned} \dot{\lambda}_{0x} &= \cosh^2\left(\lambda_{0x}\right)\left(-k_{0x}\lambda_{0x} - \Delta_{0v}\cos\left(\theta_0\right)\right)/\gamma_0, \\ \dot{\lambda}_{0y} &= \cosh^2\left(\lambda_{0y}\right)\left(-k_{0y}\lambda_{0y} - \Delta_{0v}\sin\left(\theta_0\right)\right)/\gamma_0, \end{aligned} \tag{4.7}$$

where $\Delta_{0v} = v_{0d} - v_{0s}$; k_{0x}, k_{0y} and γ_0 represent the adaptive gains; v_{0s} is the virtual linear velocity of the modular unit navigator under nested saturation.

Inspired by the pseudo-inverse feedback linearization proposed by Chwa [11], the virtual longitudinal and lateral velocities u_{10} and u_{20} (longitudinal and lateral velocity commands) of the modular unit navigator in kinematics are designed as

$$\begin{aligned} u_{10} &= -k_{10}\tanh\left(\frac{e_{10}}{k_{10}}\right) + \dot{x}_r - k_{0x}\lambda_{0x}, \\ u_{20} &= -k_{20}\tanh\left(\frac{e_{20}}{k_{20}}\right) + \dot{y}_r - k_{0y}\lambda_{0y}, \end{aligned} \tag{4.8}$$

where k_{10} and k_{20} are the control gains of the navigator.

The guidance heading (desired heading) and virtual linear velocity α_0 and v_{0d} (linear velocity command) can be expressed as

$$\begin{aligned} \alpha_0 &= \operatorname{atan2}\left(u_{20}, u_{10}\right), \\ v_{0d} &= \cos\left(\theta_0 - \alpha_0\right)\sqrt{u_{10}^2 + u_{20}^2}, \end{aligned} \tag{4.9}$$

where $\lambda_{0\theta}$ is the auxiliary variable of the navigator's heading; the corresponding adaptive auxiliary dynamic system is shown below

$$\dot{\lambda}_{0\theta} = \cosh^2\left(\lambda_{0\theta}\right)\left(-k_{0\theta}\lambda_{0\theta} - \Delta_{0w}\right)/\gamma_{0\theta}, \tag{4.10}$$

where $\Delta_{0w} = w_{0d} - w_{0s}$; $\gamma_{0\theta}$ and $k_{0\theta}$ are adaptive gains.

In order to track the virtual heading of the modular unit navigator, the virtual angular velocity w_0 (angular velocity command) is designed as

$$w_{0d} = -k_{30} \tanh\left(\frac{e_{30}}{k_{30}}\right) + \dot{\alpha}_0 - k_{0\theta}\lambda_{0\theta}, \tag{4.11}$$

where $e_{30} = \theta_0 - \alpha_0 - \gamma_{0\theta} \tanh(\lambda_{0\theta})$; k_{30} is the control gain. The formation trajectory error of the ith ($i = 1, \ldots, m$) follower is defined as

$$\begin{aligned} z_{ix} &= x_i - p_{ix} - x_0 - \gamma_i \tanh(\lambda_{ix}), \\ z_{iy} &= y_i - p_{iy} - y_0 - \gamma_i \tanh\left(\lambda_{iy}\right), \end{aligned} \tag{4.12}$$

where the auxiliary variables λ_{ix} and λ_{iy} are generated by the adaptive auxiliary dynamic system as shown below

$$\begin{aligned} \dot{\lambda}_{ix} &= \cosh^2(\lambda_{ix})(-k_{ix}\lambda_{ix} - \Delta_{iv}\cos(\theta_i))/\gamma_i, \\ \dot{\lambda}_{iy} &= \cosh^2\left(\lambda_{iy}\right)\left(-k_{iy}\lambda_{iy} - \Delta_{iv}\sin(\theta_i)\right)/\gamma_i, \end{aligned} \tag{4.13}$$

where $\Delta_{iv} = v_{id} - v_{is}$; k_{ix}, k_{iy} and γ_i are the adaptive gains of the ith follower; v_{is} is the saturated virtual linear velocity of the ith follower under nested saturation.

The cooperative position errors e_{1i} and e_{2i} of the ith ($i = 1, \ldots, m$) follower are defined as

$$\begin{aligned} e_{1i} &= \sum_{j=1}^{m} a_{ij}\left(z_{ix} - z_{jx}\right) + b_i z_{ix}, \\ e_{2i} &= \sum_{j=1}^{m} a_{ij}\left(z_{iy} - z_{jy}\right) + b_i z_{iy}. \end{aligned} \tag{4.14}$$

To ensure the convergence of the cooperative position error, the virtual longitudinal and lateral velocities u_{1i} and u_{2i} (longitudinal and lateral velocity commands) of the ith ($i = 1, \ldots, m$) follower are designed as follows

$$\begin{aligned} u_{1i} &= -k_{1i} \tanh\left(\frac{e_{1i}}{k_{1i}}\right) + \hat{h}_{ix} + \dot{p}_{ix} - k_{ix}\lambda_{ix}, \\ u_{2i} &= -k_{2i} \tanh\left(\frac{e_{2i}}{k_{2i}}\right) + \hat{h}_{iy} + \dot{p}_{iy} - k_{iy}\lambda_{iy}, \end{aligned} \tag{4.15}$$

where

$$\begin{aligned} \dot{\hat{h}}_{ix} &= -k_{ih}^{x}\left(\sum_{j=1}^{m} a_{ij}\left(\hat{h}_{ix} - \hat{h}_{jx}\right) + b_i\left(\hat{h}_{ix} - \dot{x}_0\right)\right), \\ \dot{\hat{h}}_{iy} &= -k_{ih}^{y}\left(\sum_{j=1}^{m} a_{ij}\left(\hat{h}_{iy} - \hat{h}_{jy}\right) + b_i\left(\hat{h}_{iy} - \dot{y}_0\right)\right), \end{aligned}$$

where k_{1i} and k_{2i} are the formation control gains of the ith follower, $\hat{h}_{ix}$ and $\hat{h}_{iy}$ are the estimated values of the leader's longitudinal and lateral velocities, and k_{ih}^{x} and k_{ih}^{x} are the observation gains.

Similarly, the virtual heading (reference heading) and virtual linear velocity (linear velocity command) of the ith ($i = 1, \ldots, m$) follower can be written as

$$\begin{aligned} \alpha_i &= \operatorname{atan2}(u_{2i}, u_{1i}), \\ v_{id} &= \cos(\theta_i - \alpha_i)\sqrt{u_{1i}^2 + u_{2i}^2}, \end{aligned} \tag{4.16}$$

Here, $\lambda_{i\theta}$ is the auxiliary variable of the ith follower's heading; the relative adaptive auxiliary dynamic system is shown as follows

$$\dot{\lambda}_{i\theta} = \cosh^2(\lambda_{i\theta})(-k_{i\theta}\lambda_{i\theta} - \Delta_{iw})/\gamma_{i\theta}, \tag{4.17}$$

where $\Delta_{iw} = w_{id} - w_{is}$; $\gamma_{i\theta}$ and $k_{i\theta} \in \mathbf{R}_+$ are adaptive gains.

Remark 4.1 The leader and follower adaptive auxiliary dynamic systems of Eqs. (4.7), (4.10), (4.13), and (4.17) can regulate the closed-loop control input of position trajectory and velocity through auxiliary dynamic variables when the control input is saturated, so that the control input can quickly leave the saturation region, thereby smoothing the trajectory oscillations caused by nested saturation. Usually, in adaptive design strategies, to prevent the adaptive state from being too large and causing the control input signal to be too large or the adaptive state compensation to be unreasonable, a mapping adaptive operator can be used [12] to correct the actual adaptive state. For example, with $\lambda_{i\theta}$, limit its range of change to $[\lambda_{i\theta(\min)}, \lambda_{i\theta(\max)}]$, use the mapping operator $\dot{\lambda}_{i\theta} = \operatorname{Proj}_{\lambda_{i\theta}}(\cdot)$, where $\cdot$ is the adaptive update rate, and $\operatorname{Proj}_{\lambda_{ia}}(\cdot)$ is specifically

$$\operatorname{Proj}_{\lambda_{ia}}(\cdot) = \begin{cases} 0, & \text{if } \lambda_{i\theta} \geq \lambda_{i\theta(\max)} \text{ and } \cdot > 0, \\ 0, & \text{if } \lambda_{i\theta} \leq \lambda_{i\theta(\min)} \text{ and } \cdot < 0, \\ \cdot, & \text{otherwise}. \end{cases}$$

That is, when $\lambda_{i\theta}$ exceeds the set maximum value, if there is a trend to continue increasing, i.e., $\dot{\lambda}_{i\theta} > 0$, then the value of $\lambda_{i\theta}$ remains unchanged, i.e., $\dot{\lambda}_{i\theta} = 0$; when $\lambda_{i\theta}$ exceeds the set minimum value, if there is a trend to continue decreasing, i.e., $\dot{\lambda}_{i\theta} < 0$, then the value of $\lambda_{i\theta}$ remains unchanged, i.e., $\dot{\lambda}_{i\theta} = 0$. It should be noted that the use of the mapping operator does not improve the convergence characteristics of the adaptive algorithm and the results of subsequent stability proofs. The mapping adaptive operator is also used for subsequent adaptive auxiliary dynamic systems to suppress saturation oscillations, and it ensures that the lateral/longitudinal control input is bounded.

In order to track the virtual heading of the ith follower, the virtual angular velocity (angular velocity command) is designed as follows

$$w_{id} = -k_{3i} \tanh\left(\frac{e_{3i}}{k_{3i}}\right) + \dot{\alpha}_i - k_{i\theta}\lambda_{i\theta}, \tag{4.18}$$

where $e_{3i} = \theta_i - \alpha_i - \gamma_{i\theta} \tanh(\lambda_{i\theta})$; $k_{3i} \in \mathbf{R}_+$ represents the control gain.

The position and heading dynamics of the leader and the ith follower ($i = 1, \ldots, m$) are shown as follows

$$\begin{aligned} \dot{x}_i &= v_{ie} \cos\theta_i - \Delta_{iv} \cos\theta_i + u_{1i} + \sin(\theta_i) T_i, \\ \dot{y}_i &= v_{ie} \sin\theta_i - \Delta_{iv} \sin\theta_i + u_{2i} - \cos(\theta_i) T_i, \\ \dot{\theta}_i &= w_{ie} - \Delta_{iw} + w_{id}, \end{aligned} \tag{4.19}$$

where $v_{ie} = v_i - v_{is}$ and $w_{ie} = w_i - w_{is}$ are the linear velocity and angular velocity tracking errors; $T_i \triangleq (\cos\theta_i - \cos\alpha_i) u_{2i} - (\sin\theta_i - \sin\alpha_i) u_{1i}$, satisfying

$$T_i = -2 \sin\frac{\theta_{ie}}{2}\left(u_{1i} \cos\frac{\theta_i + \alpha_i}{2} + u_{2i} \sin\frac{\theta_i + \alpha_i}{2}\right), \tag{4.20}$$

where $\theta_{ie} \triangleq \theta_i - \alpha_i$. Due to the boundedness of u_{1i} and u_{2i}, T_i is also bounded. In addition, as θ_{ie} approaches zero, T_i will also approach zero.

Then, the derivatives of the position and direction tracking errors of the leader and followers can be derived as follows

$$\begin{aligned} \dot{e}_{10} &= -k_{10} \tanh\left(\frac{e_{10}}{k_{10}}\right) + v_{0e} \cos(\theta_0) + \sin(\theta_0) T_0, \\ \dot{e}_{20} &= -k_{20} \tanh\left(\frac{e_{20}}{k_{20}}\right) + v_{0e} \sin(\theta_0) - \cos(\theta_0) T_0, \\ \dot{e}_{30} &= -k_{30} \tanh\left(\frac{e_{30}}{k_{30}}\right) + w_{0e}, \end{aligned} \tag{4.21}$$

$$\begin{aligned} \dot{z}_{ix} &= -k_{1i} \tanh\left(\frac{e_{1i}}{k_{1i}}\right) + v_{ie} \cos(\theta_i) + \sin(\theta_i) T_i + \tilde{h}_{ix}, \\ \dot{z}_{iy} &= -k_{2i} \tanh\left(\frac{e_{2i}}{k_{2i}}\right) + v_{ie} \sin(\theta_i) - \cos(\theta_i) T_i + \tilde{h}_{iy}, \\ \dot{e}_{3i} &= -k_{3i} \tanh\left(\frac{e_{3i}}{k_{3i}}\right) + w_{ie}, \end{aligned} \tag{4.22}$$

where $\tilde{h}_{ix} = \hat{h}_{ix} - \dot{x}_0$ and $\tilde{h}_{iy} = \hat{h}_{iy} - \dot{y}_0$.

The cooperative error vector of the followers can be written as $e_1 = [e_{11}, \ldots, e_{1m}]^{\mathrm{T}}$, $e_2 = [e_{21}, \ldots, e_{2m}]^{\mathrm{T}}$ and $e_3 = [e_{31}, \ldots, e_{3m}]^{\mathrm{T}}$. The derivatives of the cooperative errors $\dot{e}_1$, $\dot{e}_2$ and $\dot{e}_3$ are shown as follows

$$\begin{aligned} \dot{e}_1 &= -MK_1\xi_1 + M\bar{v}_x + MT_x + M\tilde{h}_x, \\ \dot{e}_2 &= -MK_2\xi_2 + M\bar{v}_y - MT_y + M\tilde{h}_y, \\ \dot{e}_3 &= -K_3\xi_3 + w_e, \end{aligned} \tag{4.23}$$

where

$$
\begin{aligned}
\xi_1 &= \left[\tanh\left(\frac{e_{11}}{k_{11}}\right), \ldots, \tanh\left(\frac{e_{1m}}{k_{1m}}\right)\right]^{\mathrm{T}}, \\
\xi_2 &= \left[\tanh\left(\frac{e_{21}}{k_{21}}\right), \ldots, \tanh\left(\frac{e_{2m}}{k_{2m}}\right)\right]^{\mathrm{T}}, \\
\xi_3 &= \left[\tanh\left(\frac{e_{31}}{k_{31}}\right), \ldots, \tanh\left(\frac{e_{3m}}{k_{3m}}\right)\right]^{\mathrm{T}}, \\
\bar{v}_x &= \left[v_{1e}\cos\theta_1, \ldots, v_{me}\cos\theta_m\right]^{\mathrm{T}}, \; v_e = \left[v_{1e}, \ldots, v_{me}\right]^{\mathrm{T}}, \\
\bar{v}_y &= \left[v_{1e}\sin\theta_1, \ldots, v_{me}\sin\theta_m\right]^{\mathrm{T}}, \; w_e = \left[w_{1e}, \ldots, w_{me}\right]^{\mathrm{T}}, \\
T_x &= \left[T_1\sin\theta_1, \ldots, T_m\sin\theta_m\right]^{\mathrm{T}}, \; \tilde{h}_x = \left[\tilde{h}_{1x}, \ldots, \tilde{h}_{mx}\right]^{\mathrm{T}}, \\
T_y &= \left[T_1\cos\theta_1, \ldots, T_m\cos\theta_m\right]^{\mathrm{T}}, \; \tilde{h}_y = \left[\tilde{h}_{1y}, \ldots, \tilde{h}_{my}\right]^{\mathrm{T}}, \\
K_1 &= \operatorname{diag}\{k_{11}, \ldots, k_{1m}\}, \; K_2 = \operatorname{diag}\{k_{21}, \ldots, k_{2m}\}, \\
K_3 &= \operatorname{diag}\{k_{31}, \ldots, k_{3m}\}.
\end{aligned}
$$

4.3.2 *Adaptive Saturated Extended State Observer Design*

Considering the symmetric structure of the reconfigurable swarm unit, the disturbance estimation forms of the left and right wheels are consistent. Therefore, the re-expression of the left wheel velocity dynamics of the leader and the ith follower is as follows

$$
\begin{aligned}
\dot{x}_{i1} &= I_i u_i + x_{i2}, \\
\dot{x}_{i2} &= h_i,
\end{aligned}
\tag{4.24}
$$

where $x_{i1} \triangleq \omega_{il}$, $I_i \triangleq 1/D_{11}$, $u_i \triangleq \tau_{il} - D_{12}\dot{\omega}_{ir} - f_c(\omega_{il})$, $x_{i2} \triangleq d_{il}/D_{11}$ and $h_i \triangleq \dot{d}_{il}/D_{11}$, and $|h_i| \leq \overline{h} \in \mathbf{R}_+$. The design of ASESO is as follows:

$$
\begin{aligned}
\dot{\hat{x}}_{i1} &= I_i u_i + \hat{x}_{i2} + \beta_{i1}\operatorname{sat}_{\bar{\sigma}_i}\left(x_{i1} - \hat{x}_{i1}\right), \\
\dot{\hat{x}}_{i2} &= \beta_{i2}\operatorname{sat}_{\bar{\sigma}_i}\left(x_{i1} - \hat{x}_{i1}\right), \\
\dot{\sigma}_i &= -\gamma_i\sigma_i + \vartheta_i|x_{i1} - \hat{x}_{i1}|,
\end{aligned}
\tag{4.25}
$$

Here, $\hat{x}_{i1}$ and $\hat{x}_{i2}$ represent the estimates of wheel velocity and total disturbance; $\sigma_i \in \mathbf{R}_+$ represents the adaptive saturation auxiliary dynamic variable; $\bar{\sigma}_i \triangleq \sigma_i + \epsilon_i \in \mathbf{R}_+$ is the saturation threshold, where $\epsilon_i > 0$; β_{i1} and β_{i2} are ASESO design gains; γ_i and θ_i are adaptive design gains.

The estimation error dynamics can be expressed as

$$\begin{aligned}\dot{\tilde{x}}_{i1} &= \tilde{x}_{i2} - \beta_{i1}\tilde{x}_{i1} - \beta_{i1}\delta_i, \\ \dot{\tilde{x}}_{i2} &= h_i - \beta_{i2}\tilde{x}_{i1} - \beta_{i2}\delta_i, \\ \dot{\sigma}_i &= -\gamma_i\sigma_i + \vartheta_i|\tilde{x}_{i1}|, \end{aligned} \tag{4.26}$$

where $\tilde{x}_{i1} \triangleq x_{i1} - \hat{x}_{i1}$, $\tilde{x}_{i2} \triangleq x_{i2} - \hat{x}_{i2}$ and $\delta_i \triangleq \mathrm{sat}_{\bar{\sigma}_i}(\tilde{x}_{i1}) - \tilde{x}_{i1}$.

Theorem 4.1 *Considering the reconfiguration swarm unit wheel velocity dynamics system (4.24) and the proposed adaptive saturation extended state observer (4.25). Assuming that Assumptions 4.1–4.2 are satisfied, there exist $\beta_{i1} > 0$, $\beta_{i2} > 0$, $\epsilon_i > 0$, $\gamma_i > 0$ and $\vartheta_i > 0$, making the adaptive saturation extended state observer input-state stable.*

Proof The Lyapunov function is defined as

$$V_o = \sqrt{e_o^{\mathrm{T}} P e_o} + \xi_i\sigma_i + (\xi_i + \eta_i)\max\{|\tilde{x}_{i1}| - \bar{\sigma}_i, 0\}, \tag{4.27}$$

where $e_o \triangleq \left[\tilde{x}_{i1}\ \tilde{x}_{i2}\right]^{\mathrm{T}} \in \mathbf{R}^2$ is the estimation error; $\xi_i, \eta_i \in \mathbf{R}_+$ are positive constants; $P \in \mathbf{R}^{2\times 2}$ is a positive definite matrix, satisfying $A^{\mathrm{T}}P + PA = -2I$, where $A \triangleq [-\beta_{i1}\ 1; -\beta_{i2}\ 0] \in \mathbf{R}^{2\times 2}$ is a Hurwitz matrix.

Since $\underline{\lambda}_P I \le P \le \overline{\lambda}_P I$, where $\underline{\lambda}_P \in \mathbf{R}_+$ and $\overline{\lambda}_P \in \mathbf{R}_+$ are the maximum and minimum eigenvalues of P respectively, the lower bound of V_o in Eq. (4.27) is

$$V_o \ \ge\ V_e \ \ge\ \sqrt{\underline{\lambda}_P}\|e_o\|, \tag{4.28}$$

where $V_e \triangleq \sqrt{e_o^{\mathrm{T}} P e_o}$.

Meanwhile, we have

$$|\tilde{x}_{i1}| \ \le\ l_0 V_e \ \le\ l_0\sqrt{\overline{\lambda}_P}\|e_o\|, \tag{4.29}$$

where $l_0 \triangleq 1/\sqrt{\underline{\lambda}_P} \in \mathbf{R}_+$, the upper bound of V_o is as follows:

$$\begin{aligned} V_o &\le \sqrt{\overline{\lambda}_P}\|e_o\| + (\xi_i + \eta_i)(\bar{\sigma}_i + |\tilde{x}_{i1}|) \\ &\le (\xi_i + \eta_i)\bar{\sigma}_i + \left(\sqrt{\overline{\lambda}_P} + (\xi_i + \eta_i)\, l_0\sqrt{\overline{\lambda}_P}\right)\|e_o\|. \end{aligned} \tag{4.30}$$

The Dini derivative of $|\tilde{x}_{i1}|$ is

$$\begin{aligned} D^+|\tilde{x}_{i1}| &\le l_1 V_e + \bar{\beta}_i|\delta_i| + \overline{h} \\ &\le \left(\overline{\beta}_i l_0 + l_1\right) V_e + \overline{h}, \end{aligned} \tag{4.31}$$

where $l_1 \triangleq \|A\|/\sqrt{\underline{\lambda}_P} \in \mathbf{R}_+$, $\overline{\beta}_i \triangleq \max(\beta_{i1}, \beta_{i2}) \in \mathbf{R}_+$.

Similarly, the Dini derivative of V_o [13] is

$$D^+ V_o \leq -cV_e + c_h \left(\overline{h} + \overline{\beta}_i \|\delta_i\|\right) + g_1\left(\tilde{x}_{i1}, \sigma_i\right), \tag{4.32}$$

where

$$g_1\left(\tilde{x}_{i1}, \sigma_i\right) = \begin{cases} \xi_i\left(-\gamma_i \sigma_i + \vartheta_i |\tilde{x}_{i1}|\right), & |\tilde{x}_{i1}| < \bar{\sigma}_i, \\ \eta_i\left(\gamma_i \sigma_i - \vartheta_i |\tilde{x}_{i1}|\right) + \Delta_i, & |\tilde{x}_{i1}| \geq \bar{\sigma}_i, \end{cases}$$

where $c \triangleq 1/\overline{\lambda}_P \in \mathbf{R}_+$, $c_h \triangleq \overline{\lambda}_P/\sqrt{\underline{\lambda}_P} \in \mathbf{R}_+$, $\Delta_i \triangleq (\eta_i + \xi_i)\, D^+|\tilde{x}_{i1}|$.

Choose $\vartheta_i^\star \in \mathbf{R}_+$ and $\eta_i \in \mathbf{R}_+$ as follows

$$\vartheta_i^\star = \gamma_i + \frac{c_h \overline{\beta}_i}{\eta_i},\ \eta_i = \frac{c}{2\left(l_1 + \overline{\beta}_i l_0\right)}. \tag{4.33}$$

Then, fix $\vartheta_i > \vartheta_i^\star$, and choose ξ_i to satisfy

$$\xi_i\left(\vartheta_i l_0 + l_1 + \overline{\beta}_i l_0\right) \leq \frac{c}{2}. \tag{4.34}$$

Subsequently, for some $\nu_i \in \mathbf{R}_+$, we have

$$-\vartheta_i \eta_i |\tilde{x}_{i1}| + \gamma_i \eta_i \sigma_i + c_h \overline{\beta}_i |\tilde{x}_{i1}| \leq -\nu_i \sigma_i. \tag{4.35}$$

Therefore, there exist a_{11}, a_{12}, a_{13}, a_{21}, a_{22}, and $a_{23} \in \mathbf{R}_+$ such that

$$D^+ V_o \leq g_2\left(V_e, \sigma_i\right) + g_3\left(\overline{h}\right), \tag{4.36}$$

where

$$g_2\left(V_e, \sigma_i\right) = \begin{cases} -a_{11} V_e - a_{12}\sigma_i, & |\tilde{x}_{i1}| < \bar{\sigma}_i, \\ -a_{21} V_e - a_{22}\sigma_i, & |\tilde{x}_{i1}| \geq \bar{\sigma}_i, \end{cases}$$

$$g_3\left(V_e, \sigma_i\right) = \begin{cases} a_{13}\overline{h}, & |\tilde{x}_{i1}| < \bar{\sigma}_i, \\ a_{23}\overline{h}, & |\tilde{x}_{i1}| \geq \bar{\sigma}_i, \end{cases}$$

where

$$\begin{aligned} a_{11} &\triangleq c - \xi_i \vartheta_i l_0,\ a_{12} \triangleq \xi_i \gamma_i, \\ a_{21} &\triangleq c - (\eta_i + \xi_i)\left(\overline{\beta}_i l_0 + l_1\right), \\ a_{22} &\triangleq \nu_i,\ a_{13} \triangleq c_h,\ a_{23} \triangleq c_h + \eta_i + \xi_i. \end{aligned}$$

Therefore, for some a_1 and $a_2 \in \mathbf{R}_+$, $D^+ V_o$ can satisfy

$$D^+ V_o \leq -a_1 V_o + a_2 \overline{h}. \tag{4.37}$$

According to Eqs. (4.28), (4.30), (4.31), and (4.37), the proposed ASESO is input-to-state stable relative to $\overline{h}$, and the estimation errors $\tilde{x}_{i1}$ and $\tilde{x}_{i2}$ are bounded. The final boundary analysis is shown below.

The peak phenomenon caused by the initial state deviation will trigger the saturation of ASESO, at which time $|\tilde{x}_{i1}| > \bar{\sigma}_i$. According to Eq. (4.26), the Lyapunov function candidate $V_{\tilde{x}} = \frac{1}{2}\tilde{x}_{i1}^2$ is designed to achieve convergence and the final boundary of this stage $\tilde{x}_{i1}$, where $\tilde{x}_{i2} = x_{i2} - \beta_{i2}\int_0^t \bar{\sigma}_i \text{sgn}(\tilde{x}_{i1})\, d\tau$ and

$$\text{sgn}(x) = \begin{cases} 1 & \text{if } x > 0, \\ 0 & \text{if } x = 0, \\ -1 & \text{if } x < 0. \end{cases}$$

The derivative of $\dot{V}_{\tilde{x}}$ can be expressed as

$$\begin{aligned} \dot{V}_{\tilde{x}} &= -\beta_{i1}\bar{\sigma}_i|\tilde{x}_{i1}| - \beta_{i2}|\tilde{x}_{i1}|\left(\int_0^t \sigma_i d\tau + \epsilon_i t\right) + x_{i2}\tilde{x}_{i1} \\ &\leq -\left(\beta_{i1}\bar{\sigma}_i + \beta_{i2}\epsilon_i t\right)|\tilde{x}_{i1}| + \overline{d}|\tilde{x}_{i1}|, \end{aligned} \tag{4.38}$$

where $|x_{i2}| \leq \overline{d} \in \mathbf{R}_+$.

According to Eq. (4.38), $\tilde{x}_{i1}$ converges to $\tilde{x} \leq \bar{\sigma}_i$ over time. In addition, by integrating Eq. (4.26), the upper bound of σ_i is derived as follows

$$\sigma_i \leq e^{-\gamma_i t}\sigma_i(0) + \vartheta_i\left(1 - e^{-\gamma_i t}\right) \sup_{\tau\in[0,t]} |\tilde{x}_{i1}|/\gamma_i. \tag{4.39}$$

When $|\tilde{x}_{i1}| \leq \bar{\sigma}_i$, by integrating (4.26), e_o can be obtained as

$$e_o \leq e^{-\underline{\lambda}_A t}e_o(0) + \overline{h}(1 - e^{-\overline{\lambda}_A t})/\underline{\lambda}_A, \tag{4.40}$$

where $-\underline{\lambda}_A$ and $-\overline{\lambda}_A$ are the maximum and minimum eigenvalues of $A, \overline{\lambda}_A, \underline{\lambda}_A \in \mathbf{R}_+$. At this stage, the final boundaries of $\tilde{x}_{i1}$ and $\tilde{x}_{i2}$ satisfy $|\tilde{x}_{i1}| \leq \overline{h}/\underline{\lambda}_A$. Assume $\epsilon_i > \overline{h}/\underline{\lambda}_A$ and $\gamma_i < \beta_{i1}$. Combining the above analysis, the initial peak phenomenon leads to $|\tilde{x}_{i1}| > \bar{\sigma}_i$. According to Eq. (4.38), we can adjust β_{i1}, β_{i2} and γ_i to accelerate the convergence of $\tilde{x}_{i1}$ to within $\bar{\sigma}_i$. Subsequently, the convergence of $\tilde{x}_{i1}$ will follow Eq. (4.40). At the same time, based on Eq. (4.39), a larger initial $\tilde{x}_{i1}$ will correspondingly increase σ_i, and from Eq. (4.38) it can be seen that σ_i can reduce the final boundary in Eq. (4.39) and accelerate the convergence rate. The adaptive gain γ_i of σ_i needs to be designed to balance convergence, smoothness and peak suppression, ensuring $\gamma_i < \beta_{i1}$, to ensure that $\tilde{x}_{i1}$ quickly enters the final boundary area after leaving saturation, without triggering saturation. Finally, it can be concluded that the final boundaries of $\tilde{x}_{i1}$ and $\tilde{x}_{i2}$ satisfy $|\tilde{x}_{i1}|, |\tilde{x}_{i2}| \leq \overline{h}/\underline{\lambda}_A$. □

4.3.3 Anti-disturbance Dynamic Controller Design and Stability Analysis

The control wheelbase (4.2) that integrates the wheel velocity dynamics of the leader and the ith follower can be expressed as:

$$\begin{aligned}\tau_{ir} &= D_{11}\left(\dot{\omega}_{dir} + k_{i\omega}\tilde{\omega}_{ir}\right) + D_{12}\dot{\omega}_{il} + f_c\left(\omega_{ir}\right) + D_{11}\hat{d}_{ir},\\ \tau_{il} &= D_{21}\left(\dot{\omega}_{dil} + k_{i\omega}\tilde{\omega}_{il}\right) + D_{22}\dot{\omega}_{ir} + f_c\left(\omega_{il}\right) + D_{22}\hat{d}_{il},\end{aligned} \tag{4.41}$$

where $k_{i\omega} \in \mathbf{R}_+$ is the control gain; ω_{dir} and ω_{dil} are the desired velocities; $\tilde{\omega}_{ir} = \omega_{dir} - \omega_{ir}$ and $\tilde{\omega}_{il} = \omega_{dil} - \omega_{il}$ are the wheel velocity tracking errors; $\hat{d}_{ir}$, $\hat{d}_{il}$, $\tilde{d}_{ir}$ and $\tilde{d}_{il}$ are the total disturbance estimates and their errors from the adaptive saturation extended observer.

The error dynamics of the wheel velocity dynamics are shown as follows

$$\begin{aligned}\dot{\tilde{\omega}}_{ir} &= -k_{i\omega}\tilde{\omega}_{ir} + \tilde{d}_{ir},\\ \dot{\tilde{\omega}}_{il} &= -k_{i\omega}\tilde{\omega}_{il} + \tilde{d}_{il}.\end{aligned} \tag{4.42}$$

The corresponding linear velocity and angular velocity tracking errors and their dynamics can be expressed as

$$\begin{aligned}v_{ie} &= \frac{r_i\left(\tilde{\omega}_{ir} + \tilde{\omega}_{il}\right)}{2}, \ \dot{v}_{ie} = -k_{iv}v_{ie} + \delta_{iv},\\ w_{ie} &= \frac{r_i\left(\tilde{\omega}_{ir} - \tilde{\omega}_{il}\right)}{2b_i}, \ \dot{w}_{ie} = -k_{iw}w_{ie} + \delta_{iw},\end{aligned} \tag{4.43}$$

Among them, $\delta_{iv} = \frac{r_i(\tilde{d}_{ir}+\tilde{d}_{il})}{2}$, $k_{iv} = k_{i\omega}$, $\delta_{iw} = \frac{r_i(\tilde{d}_{ir}-\tilde{d}_{il})}{2b_i}$ and $k_{iw} = k_{i\omega}$. The related matrices and vectors are defined as $K_v = \text{diag}\{k_{1v}, \ldots, k_{mv}\}$, $K_w = \text{diag}\{k_{1w}, \ldots, k_{mw}\}$, $\delta_v = [\delta_{1v}, \ldots, \delta_{mv}]^{\mathrm{T}}$ and $\delta_w = [\delta_{1w}, \ldots, \delta_{mw}]^{\mathrm{T}}$. Because the disturbance estimation errors of the leader and followers in the reconfigurable swarm unit modality are bounded, there exist constants $\bar{\delta}_{0v}$, $\bar{\delta}_{0w}$, $\bar{\delta}_v$ and $\bar{\delta}_w \in \mathbf{R}_+$, such that $|\delta_{0v}| \leq \bar{\delta}_{0v}$, $|\delta_{0w}| \leq \bar{\delta}_{0w}$, $|\delta_v| \leq \bar{\delta}_v$ and $|\delta_v| \leq \bar{\delta}_w$.

Theorem 4.2 *Considering the reconfigurable swarm system of swarm units based on the leader-follower structure (3.4)–(4.2), subject to velocity and acceleration nested saturation constraints (4.4). Assuming that Assumptions 4.1–4.3 are satisfied, then the distributed formation leader controller (4.8), (4.9) and (4.11) and the follower controller (4.15), (4.16) and (4.18), combined with the smooth saturation oscillating leader and follower adaptive anti-saturation system (4.7), (4.10), (4.13) and (4.17) and the lower layer fusion adaptive saturation extended state observer (4.25) dynamic feedback controller (4.41), can ensure that the motion control errors of the formation leader and followers and their corresponding dynamic control errors,* e_{10}, e_{20}, e_{30}, e_1, e_2, e_3, v_{0e}, w_{0e}, v_e *and* w_e *are ultimately bounded.*

Proof The Lyapunov candidate function of the leader is as follows

$$V_l = k_{10}\ln\left(\cosh\left(\frac{e_{10}}{k_{10}}\right)\right) + k_{20}\ln\left(\cosh\left(\frac{e_{20}}{k_{20}}\right)\right) + k_{30}\ln\left(\cosh\left(\frac{e_{30}}{k_{30}}\right)\right) + \frac{1}{2}v_{0e}^2 + \frac{1}{2}w_{0e}^2. \tag{4.44}$$

The derivative of V_l is as follows

$$\begin{aligned}\dot{V}_l &= -k_{10}\xi_{10}^2 - k_{20}\xi_{20}^2 - k_{30}\xi_{30}^2 - k_{0v}v_e^2 - k_{0w}w_e^2 + w_e\delta_{0w} + v_e\delta_{0v} \\ &\quad + \xi_{10}\sin(\theta_0)T_0 - \xi_{20}\cos(\theta_0)T_0 + \xi_{10}v_{0e}\cos(\theta_0) + \xi_{20}v_{0e}\sin(\theta_0) + \xi_{30}w_{0e} \\ &\leq -\bar{k}_{10}\xi_{10}^2 - \bar{k}_{20}\xi_{20}^2 - \bar{k}_{30}\xi_{30}^2 - \bar{k}_{0v}v_e^2 - \bar{k}_{0w}w_e^2 + \bar{\delta}_0,\end{aligned} \tag{4.45}$$

where $\xi_{10} = \tanh(e_{10}/k_{10})$, $\xi_{20} = \tanh(e_{20}/k_{20})$, $\xi_{30} = \tanh(e_{30}/k_{30})$; $\bar{k}_{10} > \frac{3}{4}$, $\bar{k}_{20} > \frac{3}{4}$, $\bar{k}_{30} > \frac{1}{2}$, $\bar{k}_{0v} > \frac{5}{4}$, $\bar{k}_{0w} > \frac{3}{4}$; $\bar{\delta}_0 = 2T_0^2 + \bar{\delta}_v^2 + \bar{\delta}_w^2 \in \mathbf{R}_+$. According to Eqs. (4.44) and (4.45), the position tracking errors ξ_{10}, ξ_{20}, ξ_{30}, v_{0e} and w_{0e} of the leader in the swarm formation can be reconstructed and are ultimately bounded.

The Lyapunov function of the leader estimator error is defined as

$$V_h = \frac{1}{2}\tilde{h}_x^{\mathrm{T}}\mathcal{M}\tilde{h}_x + \frac{1}{2}\tilde{h}_y^{\mathrm{T}}\mathcal{M}\tilde{h}_y, \tag{4.46}$$

where $e_h = \mathrm{col}\{\tilde{h}_x, \tilde{h}_y\}$; $\bar{X}_h \triangleq [\dot{x}_0, \ldots, \dot{x}_0]^{\mathrm{T}} \in \mathbf{R}^m$ and $\bar{Y}_h = [\dot{y}_0, \ldots, \dot{y}_0]^{\mathrm{T}} \in \mathbf{R}^m$. $\bar{X}_h$ and $\bar{Y}_h$ are bounded, and there exists a constant δ_h such that $\|\mathrm{col}\{\bar{X}_h, \bar{X}_h\}\| \leq \delta_h \in \mathbf{R}_+$.

The derivative of V_h is as follows

$$\begin{aligned}\dot{V}_h &= -\tilde{h}_x^{\mathrm{T}}\mathcal{M}\left(K_h^x\mathcal{M}\tilde{h}_x + \bar{X}_0\right) - \tilde{h}_y^{\mathrm{T}}\mathcal{M}\left(K_h^y\mathcal{M}\tilde{h}_y + \bar{Y}_0\right) \\ &\leq -\frac{\lambda_{\min}(Q_h)}{\lambda_{\max}(P_h)}V_h + 2\bar{\delta}_h\sqrt{V_h}\sqrt{P_h},\end{aligned} \tag{4.47}$$

where

$$\begin{aligned}K_h^x &= \mathrm{diag}\{k_{1h}^x, \ldots, k_{mh}^x\},\ K_h^y = \mathrm{diag}\{k_{1h}^y, \ldots, k_{mh}^y\}, \\ Q_h &= \mathrm{diag}\{\mathcal{M}K_h^x\mathcal{M}, \mathcal{M}K_h^y\mathcal{M}\},\ P_h = \mathrm{diag}\{0.5\mathcal{M}, 0.5\mathcal{M}\},\end{aligned}$$

where K_h^x, K_h^y, Q_h and P_h are positive definite matrices.

There exists a set $\Omega_h = \{e_h \in \mathbf{R}^{2m} | V_h \leq \frac{4\bar{\delta}_h^2\|P_h\|\lambda_{\max}^2(P_h)}{\lambda_{\min}^2(Q_h)}\}$, such that for $e_h \in \{\mathbf{R}^{2m}\backslash\Omega_h\}$, $V_h \leq 0$, resulting in e_h converging to Ω_h. Therefore, there exist constants $\bar{\delta}_{hx}, \bar{\delta}_{hy} \in \mathbf{R}_+$ such that $\|\tilde{h}_x\| \leq \bar{\delta}_{hx}$, $\|\tilde{h}_y\| \leq \bar{\delta}_{hy}$.

The Lyapunov candidate function for the follower is chosen as

$$V_f = \sum_{i=1}^{m} k_{1i} \ln\left(\cosh\left(\frac{e_{1i}}{k_{1i}}\right)\right) + \sum_{i=1}^{m} k_{2i} \ln\left(\cosh\left(\frac{e_{2i}}{k_{2i}}\right)\right) + \sum_{i=1}^{m} k_{3i} \ln\left(\cosh\left(\frac{e_{3i}}{k_{3i}}\right)\right) + \frac{1}{2} v_e^{\mathrm{T}} v_e + \frac{1}{2} w_e^{\mathrm{T}} w_e. \tag{4.48}$$

The derivative of V_f is as follows

$$\dot{V}_f = -\xi_1^{\mathrm{T}} \mathcal{M} K_1 \xi_1 - \xi_2^{\mathrm{T}} \mathcal{M} K_2 \xi_2 - \xi_3^{\mathrm{T}} K_3 \xi_3 + v_e^{\mathrm{T}} \delta_v + w_e^{\mathrm{T}} \delta_w + \xi_1^{\mathrm{T}} \mathcal{M} \bar{v}_x + \xi_1^{\mathrm{T}} \mathcal{M} T_x + \xi_2^{\mathrm{T}} \mathcal{M} \bar{v}_y - \xi_2^{\mathrm{T}} \mathcal{M} T_y + \xi_1^{\mathrm{T}} \mathcal{M} \tilde{h}_x + \xi_2^{\mathrm{T}} \mathcal{M} \tilde{h}_y + \xi_3^{\mathrm{T}} w_e - v_e^{\mathrm{T}} K_v v_e - w_e^{\mathrm{T}} K_w w_e. \tag{4.49}$$

The derivative of V_f can satisfy the following inequality

$$\dot{V}_f \le -\bar{K}_1 \|\xi_1\|^2 - \bar{K}_2 \|\xi_2\|^2 - \bar{K}_3 \|\xi_3\|^2 - \bar{K}_v \|v_e\|^2 - \bar{K}_w \|w_e\|^2 + \Delta_\delta + \Delta_h + \Delta_T, \tag{4.50}$$

where

$$\begin{aligned}
\bar{K}_1 &= \lambda_{\min}\{\mathcal{M} K_1\} - \frac{1}{2} - \frac{1}{2}\lambda_{\max}\{\mathcal{M}\} \in \mathbf{R}_+, \\
\bar{K}_2 &= \lambda_{\min}\{\mathcal{M} K_2\} - \frac{1}{2} - \frac{1}{2}\lambda_{\max}\{\mathcal{M}\} \in \mathbf{R}_+, \\
\bar{K}_3 &= \lambda_{\min}\{K_3\} - \frac{1}{2} \in \mathbf{R}_+, \\
\bar{K}_v &= \lambda_{\min}\{K_v\} - \frac{5}{4} \in \mathbf{R}_+, \\
\Delta_h &= \lambda_{\max}\{\mathcal{M}\}\bar{\delta}_{hx}^2 + \lambda_{\max}\{\mathcal{M}\}\bar{\delta}_{hy}^2 \in \mathbf{R}_+, \\
\Delta_T &= \lambda_{\max}\{\mathcal{M}\}\|T_x\|^2 + \lambda_{\max}\{\mathcal{M}\}\|T_y\|^2 \in \mathbf{R}_+, \\
\bar{K}_w &= \lambda_{\min}\{K_w\} - \frac{3}{4} \in \mathbf{R}_+, \\
\Delta_\delta &= \|\bar{\delta}_v\|^2 + \|\bar{\delta}_w\|^2 \in \mathbf{R}_+.
\end{aligned}$$

Based on Eqs. (4.44), (4.45), (4.48) and (4.50), the error of the formation control system based on the leader-follower structure is ultimately bounded, and the corresponding error bounds of the leader and followers can be adjusted by adjusting the gains of each subsystem.

4.4 Simulation and Experimental Verification

4.4.1 Simulation Result Analysis

The simulation of the reconfigurable swarm unit formation under nested saturation was implemented in MATLAB R2021b, with a simulation duration of 40 s and a control cycle of 1 ms. The entire unit formation control framework uses a leader-follower structure. In the simulation, there are three units in total, one of which serves as the leader and the other two as followers. Their communication topology structure is shown as follows:

$$\mathcal{A} = \begin{bmatrix} 0 & 1 \\ 1 & 0 \end{bmatrix}, \mathcal{L} = \begin{bmatrix} 1 & -1 \\ -1 & 1 \end{bmatrix}, \mathcal{B} = \begin{bmatrix} 1 & 0 \\ 0 & 0 \end{bmatrix}.$$

The linear velocity and angular velocity of the leading navigator are set to $v_r = 0.2\ \mathrm{m/s}$ and $w_r = 0.1\ \mathrm{rad/s}$, respectively. The shape configuration of the unit formation is $\left[p_{1x}, p_{1y}, p_{2x}, p_{2y}\right] = \varrho\left[-1.0, 0.0, -0.5, \sqrt{3}/2\right]$ m, where ϱ is the shape scaling factor. The initial motion states of the leading navigator and the following modular units in the reconfigurable swarm are preset as

$$\begin{aligned} [x_0, y_0, \theta_0]^{\mathrm{T}} &= [1\ \mathrm{m}, 1\ \mathrm{m}, 0\ \mathrm{rad}]^{\mathrm{T}}, \\ [x_1, y_1, \theta_1]^{\mathrm{T}} &= [-1\ \mathrm{m}, 1\ \mathrm{m}, 0\ \mathrm{rad}]^{\mathrm{T}}, \\ [x_2, y_2, \theta_2]^{\mathrm{T}} &= [-1\ \mathrm{m}, -1\ \mathrm{m}, 0\ \mathrm{rad}]^{\mathrm{T}}. \end{aligned}$$

The initial wheel velocity of the modular unit swarm is set to 0. The thresholds for the linear velocity, angular velocity, and acceleration of each modular unit are set to 1 m/s, 1 rad/s, 0.4 $\mathrm{m/s^2}$, and 0.4 $\mathrm{rad/s^2}$, respectively. The unknown external random disturbances applied to the wheel velocity systems of the leading navigator and the followers are chosen as $d_{il} = 10\sin(t) + 2\cos(0.6t)\ \mathrm{N\cdot m}$ and $d_{ir} = 10\cos(t) + 2\sin(0.6t)\ \mathrm{N\cdot m}$.

The formation control method based on error feedback is prone to trigger the saturation of the velocity or acceleration of the reconfigurable swarm formation units due to large initial errors or sudden changes in the formation. Therefore, these two scenarios are introduced to highlight the anti-saturation advantage of the proposed method. Hence, from 0 s to 20 s, the scaling factor is set to 1, and from 20 s to 40 s, the scaling factor is set to 2. To demonstrate the anti-saturation effect in the simulation, all compared unit controllers in the dynamics layer use ASESO, while the proposed method or a variant of the proposed method without adaptive anti-saturation is used in the kinematics layer, labeled as Proposed and ProposedWOAW respectively. The kinematics layer also uses the formation control scheme based on the leader-following structure proposed by Chang et al.[14] and the formation control method combining this scheme with adaptive anti-saturation, labeled as Chang2022 and Chang2022AW respectively. Moreover, to highlight the advantage of the proposed control scheme in suppressing the peak estimation of unknown disturbances in the initial stage in the simulation, all compared kinematics layers use the proposed method, while the dynamics layer uses ASESO and traditional LESO (Linear Extended State Observer) as disturbance estimation schemes. To ensure fairness, the initial values and observation gains of the two estimators are set the same. To ensure the fairness of the simulation comparison, all compared upper kinematics layers use equal gains, specifically $k_{10} = k_{11} = k_{12} = k_{20} = k_{21} = k_{22} = 1.0$, $k_{30} = k_{31} = k_{32} = 3.0$, $k_{1h}^{x} = k_{1h}^{y} = k_{2h}^{x} = k_{2h}^{y} = 100$, and $\lambda_{0x} = \lambda_{1x} = \lambda_{2x} = \lambda_{0y} = \lambda_{1y} = \lambda_{2y} = \lambda_{0\theta} = \lambda_{1\theta} = \lambda_{2\theta} = 12$. The gains of the dynamics controller are $k_{0\omega} = k_{1\omega} = k_{2\omega} = 10$. The gains of ASESO and LESO are set to $\beta_{i1} = 1000$, $\beta_{i2} = 200000$, $\gamma_i = 5$, $\vartheta_i = 0.2$ and $\epsilon_i = 0.01$ (ASESO), and $\beta_{i1} = 1000$ and $\beta_{i2} = 200000$ (LESO).

(1) Kinematic Self-Adaptive Anti-Saturation: Fig. 4.2a–d clearly demonstrates the comparative effects of the modular unit on formation and coordination errors. As can be seen from Fig. 4.2a, c, in the Proposed scheme, the three modular units accurately track the expected formation at 8s, 16s, 24s, and 32s, while in the ProposedWOAW scheme, except at 16s, the three modular units are unable to maintain the expected formation at other times. Figure 4.2b, d show that in the initial stage and during the formation transition, the oscillation amplitude and convergence time of the coordination error in the proposed method are generally superior to the Proposed-WOAW scheme. Therefore, the proposed method has obvious advantages in terms of smoothness and anti-saturation. Figure 4.2a–d shows the advantages of the proposed self-adaptive anti-saturation method in oscillation suppression and trajectory smoothing compared to other formation control methods. In addition, the performance of the proposed formation method is generally not inferior to the effects of the Chang2022AW formation method.

(2) Suppression of Kinematic Disturbance: Fig. 4.2e–f show the comparison results of disturbance estimation in the ASESO and LESO of the modular unit wheel velocity kinematic controller. In Fig. 4.2e, when using ASESO, the peak values of the disturbance estimation of the three modular units' left wheels are 40, 46, and 47, respectively, while when using LESO, the peak values are 84, 116, and 116, respectively. In Fig. 4.2f, when using ASESO, the peak values of the disturbance estimation of the three modular units' right wheels are 38, 43, and 43, respectively, while when using LESO, the peak values are 100, 116, and 109, respectively. The above comparison shows that the proposed ASESO can effectively suppress the initial estimation peak, and too high peak values can easily lead to excessive motor torque current, thereby triggering severe oscillations and loss of control. Therefore, in practical applications, the peak suppression capability of ASESO means that higher disturbance observer gains can be introduced, thereby achieving higher precision and stronger robustness of disturbance estimation.

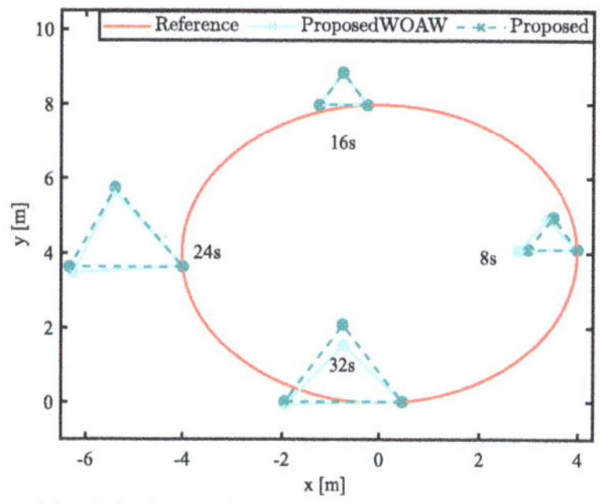

(a) Unit formation trajectories in Proposed

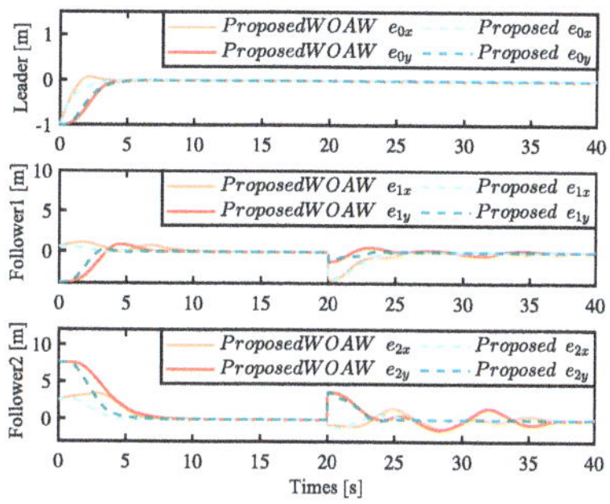

(b) Unit formation errors in Proposed

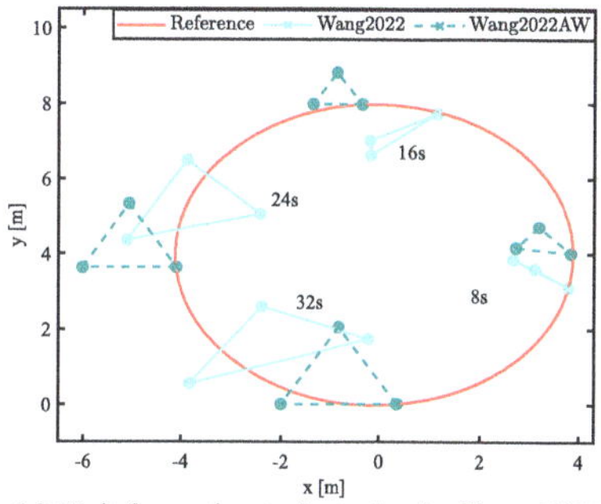

(c) Unit formation trajectories in Chang2020

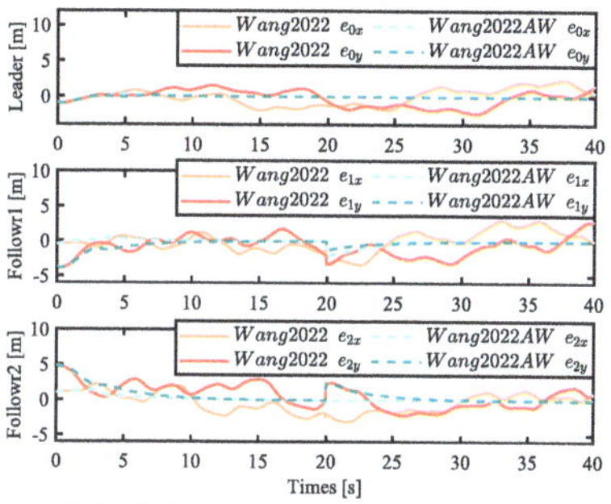

(d) Unit formation errors in Chang2020

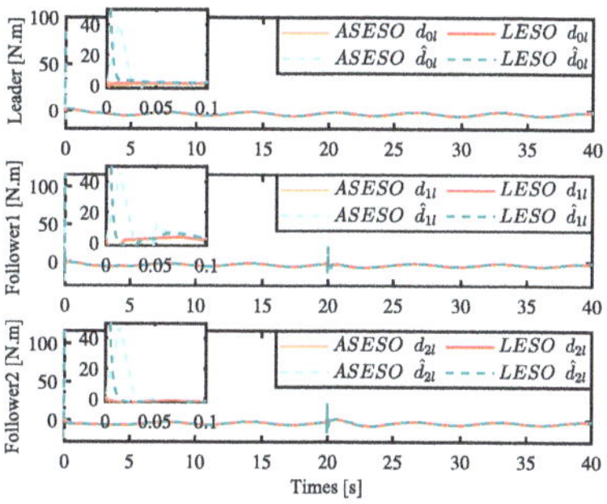

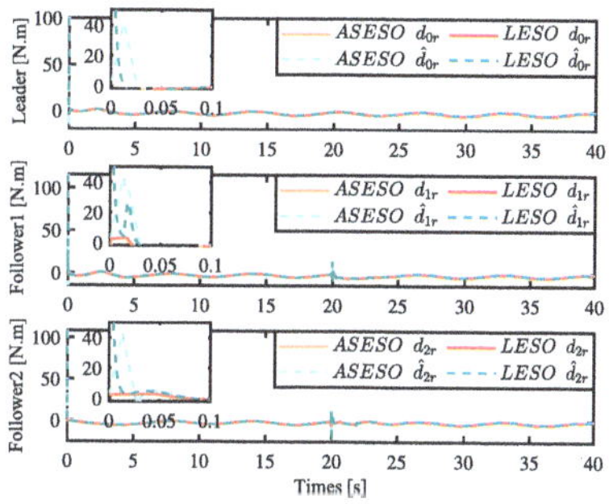

(e) Left wheel disturbance estimation comparison (f) Right wheel disturbance estimation comparison

Fig. 4.2 Comparison of formation trajectory and velocity tracking of unit modality formation

4.4.2 Experimental Results Analysis

The experimental investigations of reconfigurable swarm unit formation were implemented using the testbed developed in Chap. 2. The platform consists of three modular units, with a movement area of 2.5 m × 2.5 m. The communication topology structure of the modular unit formation is consistent with the simulation. Each unit is equipped with an STM32 microcontroller, two torque motors with wheel velocity and current feedback, and a WiFi device for communication with the host, which can be used for low-level dynamic control. The control cycle is 2 ms. The C code of the low-level dynamic controller is programmed into the microcontroller for managing the wheel velocity control of the modular unit and implementing bidirectional communication with the host. The high-level kinematic formation control with a control cycle of 20 ms is written in C++ code and runs as a ROS node on the host. The host ROS communicates with the TCP communication node and the Apriltag visual positioning node, respectively sending linear velocity and angular velocity commands to the modular unit microcontroller and obtaining the real-time position of the Tag placed on top of the modular unit. Since the host can obtain the status of each modular unit and send them velocity commands, it can simulate a distributed control architecture to verify the effectiveness of the proposed method.

Four scenarios are used to evaluate the anti-saturation and disturbance suppression performance of the nested saturation coding queue, specifically as follows: In the first two scenarios, under no-load conditions, the lower-level dynamics controller integrates ASESO, and the upper-level kinematics controllers respectively adopt the proposed method and the proposed method without adaptive anti-saturation (for verifying the anti-saturation effect). In the subsequent two scenarios, under load conditions, the lower-level dynamics controllers respectively integrate ASESO and LESO (for verifying the disturbance suppression effect), while the upper-level kinematics controllers all adopt the proposed method. These four scenarios are respectively marked as C1, C2, C3, and C4. To ensure the fairness of the comparison, the upper-level kinematics controllers used for comparison adopt the same gains, set as $k_{10} = k_{11} = k_{12} = k_{20} = k_{21} = k_{22} = 1.2, k_{30} = k_{31} = k_{32} = 9.0$, $k^x_{1h} = k^y_{1h} = k^x_{2h} = k^y_{2h} = 50$, $\lambda_{0x} = \lambda_{1x} = \lambda_{2x} = \lambda_{0y} = \lambda_{1y} = \lambda_{2y} = 30$ and $\lambda_{0\theta} = \lambda_{1\theta} = \lambda_{2\theta} = 10$, while the dynamics controller gains are $k_{0\omega} = k_{1\omega} = k_{2\omega} = 1.2$. In order to maximize the robustness of disturbance estimation, the gains of ASESO and LESO are chosen to be the maximum allowable values under the condition of avoiding severe oscillation of the torque motor, specifically $\beta_{i1} = 1000$, $\beta_{i2} = 250000$ (ASESO) and $\beta_{i1} = 140$, $\beta_{i2} = 4900$ (LESO).

(1) Kinematic Adaptive Anti-Saturation: Fig. 4.3a–c shows the comparison of the anti-saturation effects of C1 and C2 under no-load disturbance. In Fig. 4.3a, the smoothness and integrity of the trajectory of C1 are significantly better than those of C2. Figure 4.3a, b show that in the initial stage and the formation transition stage, the convergence time and oscillation amplitude of C1 are also significantly better than those of C2, but the steady-state accuracy of C1 (0.041 m) and C2 (0.057 m) are basically equivalent. As can be seen from Fig. 4.3c, the oscillations of the saturation linear velocity and angular velocity commands of C1 are also much milder than those of C2. Overall, the proposed method can effectively smooth the formation trajectory and suppress saturation oscillations under nested velocity and acceleration movements.

(2) Suppression of Kinematic Disturbances: Fig. 4.3d–f show the comparative control results of C3 and C4 in kinematics and dynamics under unknown load disturbance. In Fig. 4.3d, the trajectory smoothness and shape integrity of C3 are slightly better than C4, and can almost rival C1 (as shown in Fig. 4.3a). Figure 4.3a, b and

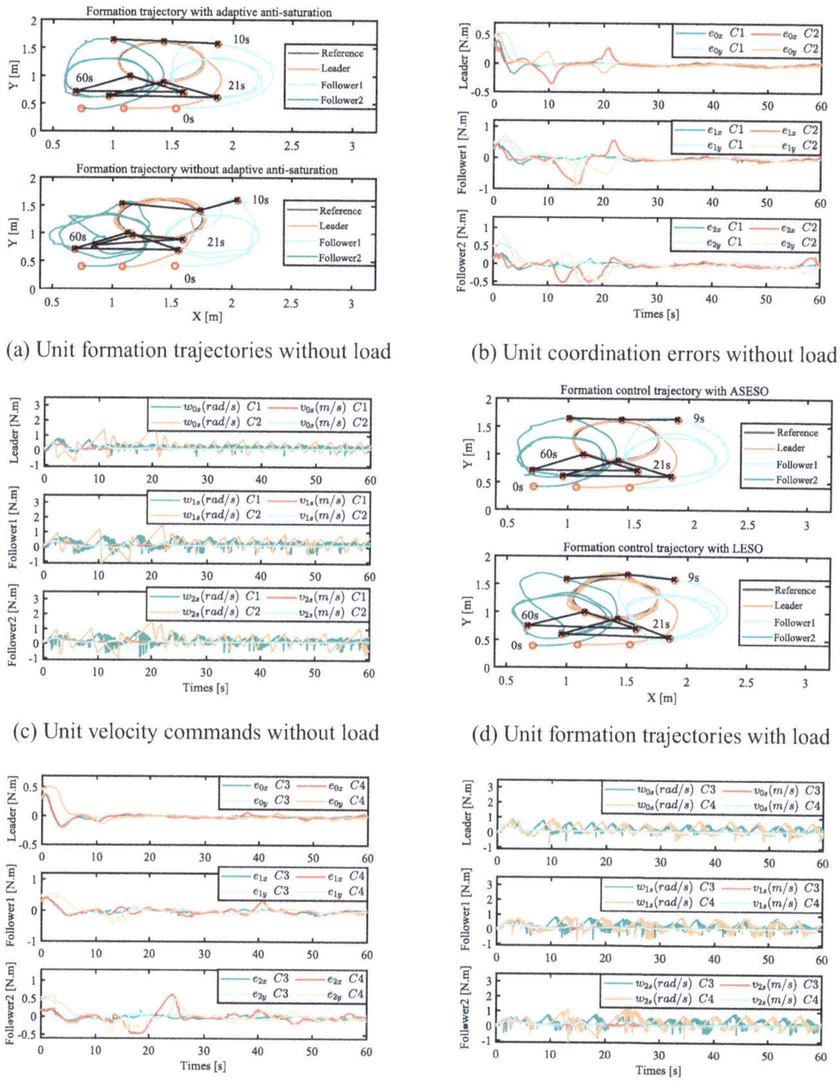

(a) Unit formation trajectories without load

(b) Unit coordination errors without load

(c) Unit velocity commands without load

(d) Unit formation trajectories with load

(e) Unit coordination errors with load

(f) Unit velocity commands with load

Fig. 4.3 Experimental results of formation control

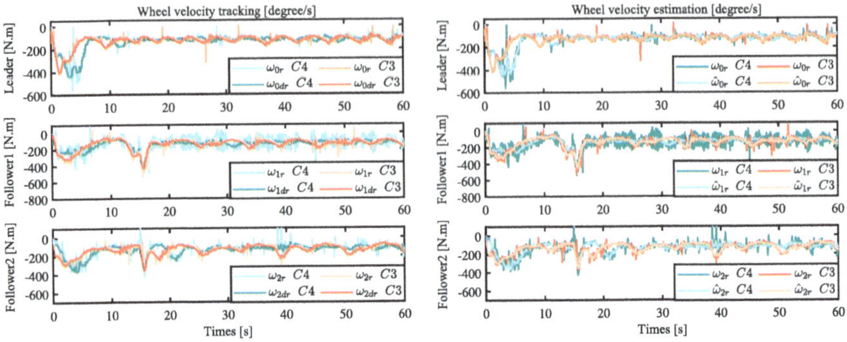

(a) Wheel velocity tracking with load (b) Wheel velocity estimation with load

Fig. 4.4 The comparison of unit modality dynamic control performance with load

d, e show that C3 also outperforms C4 in terms of convergence time and oscillation amplitude in the initial and transition stages, and its transient performance in these stages is not inferior to C1. Meanwhile, the steady-state accuracy of C1 (0.041m) and C3 (0.055m) are almost equivalent, but significantly better than C4 (0.089m). This chapter adopts a kinematic-dynamic cascaded control structure, where the lower layer dynamic uncertainty can be suppressed through feedforward compensation of disturbance estimation. In practical applications, ASESO allows the use of higher observer gains relative to LESO, thereby achieving higher precision and robustness in disturbance estimation and suppression, as clearly demonstrated in Fig. 4.4a–b through wheel velocity tracking, estimation, and disturbance estimation. In summary, the proposed method achieves superior wheel velocity tracking performance and kinematic coordination control performance in the cascaded and feedforward compensation structure, and has stronger adaptability in complex environments.

4.5 Summary

This chapter addresses the model constraints, velocity and acceleration nested saturation, and other limiting constraints and uncertainties within and outside the model that exist in multi-unit formation control. A distributed formation framework and its constraint and disturbance-resistant adaptive enhancement strategy are designed. First, a distributed controller is designed using a position-attitude-velocity cascaded control structure, realizing the basic lead-following and formation maintenance problems of the agent. Secondly, combining the kinematic level with non-integral constraints and nested saturation, a distributed adaptive auxiliary dynamic system is designed under the distributed topology to suppress possible triggered saturation oscillations. At the kinematic level, ASESO is introduced to obtain low peak disturbance estimation and its feedforward compensation signal, to enhance the robustness of the agent control in practical scenarios. Finally, trajectories that are prone to trigger saturation and large load disturbances are introduced into the agent swarm simulation and experiment respectively, highlighting the adaptive enhancement effect of the proposed method in anti-saturation and anti-disturbance.

References

1. Liu, W., Wang, X., Li, S.: Formation control for leader-follower wheeled mobile robots based on embedded control technique. IEEE Trans. Control Syst. Technol. **31**(1), 265–280 (2023)
2. Sarrafan, N., Zarei, J.: Bounded observer-based consensus algorithm for robust finite-time tracking control of multiple nonholonomic chained-form systems. IEEE Trans. Autom. Control **66**(10), 4933–4938 (2021)
3. Wang, Y., Wang, D., Yang, S., Shan, M.: A practical leader-follower tracking control scheme for multiple nonholonomic mobile robots in unknown obstacle environments. IEEE Trans. Control Syst. Technol. **27**(4), 1685–1693 (2019)
4. Liu, A., Zhang, W., Yu, L., Yan, H., Zhang, R.: Formation control of multiple mobile robots incorporating an extended state observer and distributed model predictive approach. IEEE Trans. Syst. Man Cybern. Syst. **50**(11), 4587–4597 (2020)
5. Li, L., Kuang, C., Xia, Y., Qiang, J.: Formation control of nonholonomic mobile robots with inaccurate global positions and velocities. Int. J. Robust Nonlinear Control **32**(18), 9776–9790 (2022)
6. Tzafestas, S.: Introduction to Mobile Robot Control. Elsevier, Amsterdam (2013)
7. Yao, J., Deng, W., Jiao, Z.: Adaptive control of hydraulic actuators with luGre model-based friction compensation. IEEE Trans. Ind. Electron. **62**(10), 6469–6477 (2015)
8. Rad, S., Khosravi, A., Sarhadi, P.: Pitch autopilot design for an autonomous aerial vehicle in the presence of amplitude and rate saturation. Aerosp. Sci. Technol. **108**, 106371 (2021)
9. Lashkari, N., Biglarbegian, M., Yang, S.X.: Development of a novel robust control method for formation of heterogeneous multiple mobile robots with autonomous docking capability. IEEE Trans. Autom. Sci. Eng. **17**(4), 1759–1776 (2020)
10. Li, S., Wang, X.: Finite-time consensus and collision avoidance control algorithms for multiple AUVs. Automatica **49**, 3359–3367 (2013)
11. Chwa, D.: Tracking control of differential-drive wheeled mobile robots using a backstepping-like feedback linearization. IEEE Trans. Syst. Man Cybern. A **40**(6), 1285–1295 (2010)
12. Xu, L., Yao, B.: Adaptive robust control of mechanical systems with non-linear dynamic friction compensation. Int. J. Control **81**(2), 167 176 (2008)
13. Shakarami, M., Esfandiari, K., Suratgar, A.A., Talebi, H.A.: Peaking attenuation of high-gain observers using adaptive techniques: State estimation and feedback control. IEEE Trans. Autom. Control **65**(10), 4215–4229 (2020)
14. Chang, S., Wang, Y., Zuo, Z., Yang, H.: Fixed-time formation control for wheeled mobile robots with prescribed performance. IEEE Trans. Control Syst. Technol. **30**(2), 844–851 (2022)

Chapter 5
Unit Modality Coordination Control Under Environmental Constraints

Abstract The distributed cooperative control of reconfigurable swarm in unit modalities under nested motion saturation and environmental safety constraints is addressed in this chapter. An adaptive smooth safety-cooperative method is proposed. By incorporating dynamic safety barriers and low-trigger potential fields, formation safety is ensured in obstacle-rich environments. An adaptive auxiliary dynamic system is introduced to suppress saturation-induced oscillations. A distributed coordination controller is designed, and both simulations and experiments validate the safety and robustness of the proposed method.

Keywords Distributed coordination control · Safety barrier · Artificial potential field · Nested motion saturation · Adaptive oscillation suppression

5.1 Introduction

This chapter further adds the requirement of environmental safety constraints based on the nested saturation constraints in Chap. 4, and studies the performance adaptability problem of the modular unit swarm under the existence of obstacle avoidance and collision avoidance constraints. Most of the existing mobile modular unit safety coordination adopts the reaction strategy of repulsive force avoidance, which easily triggers the saturation oscillation of the formation trajectory of the interconnected modular units in the network and the motion delay of their task game when the repulsive force is triggered, even causing the overall loss of control of the modular unit coordination. This chapter proposes a framework for smooth formation and safety assurance of distributed unit swarm, addressing the triggering constraints such as environmental safety constraints and nested motion saturation that exist in unit formation. Firstly, a dynamic safety barrier and a low-trigger potential mixed safety strategy are introduced. The safety distance between the unit and adjacent formation members is ensured through a low-trigger repulsion field, and a dynamic safety barrier is used to generate an acceleration envelope to guide the unit around non-cooperative obstacles. The combined repulsion field ensures more comprehensive safety. Compared to traditional repulsion strategies [1–3], it is easier to ensure

T. Jiang et al., *Multi-modality Control of Reconfigurable Swarm*,
https://doi.org/10.1007/978-981-95-0305-6_5

smooth trajectory characteristics under multi-task coordination. Then, a composite adaptive auxiliary dynamic system is designed and integrated into the distributed formation control feedback structure to suppress the oscillation of the formation trajectory caused by the inherent nested motion saturation of the unit and the safety acceleration envelope. Compared to the moving unit safety formation method [3, 4], it considers the negative impact of the actual existing velocity and acceleration nested saturation constraints on safety coordination. Finally, a reconfigurable swarm unit safety coordination numerical simulation and experimental platform is established. By comparing with existing safety formation methods, the proposed method is validated to enhance the adaptability of safety and smoothness performance in unit formation.

5.2 Problem Description

The ith modular unit follower considers the kinematic model (3.4), taking into account the kinematic model of the modular unit wheelbase center relative to the measurement position deviation as

$$\begin{cases} \dot{x}_{il} &= v_i \cos(\theta_i) - l_i w_i \sin(\theta_i), \\ \dot{y}_{il} &= v_i \sin(\theta_i) + l_i w_i \cos(\theta_i), \\ \dot{\theta}_{il} &= w_i, \end{cases} \tag{5.1}$$

where x_i, y_i and θ_i are the center position and heading of the ith follower; $x_{il} = x_i + l_i \cos(\theta_i)$, $y_{il} = y_i + l_i \sin(\theta_i)$ and $\theta_{il} = \theta_i$ are the position and heading of the measurement point of the ith follower, l_i is the deviation; v_i and w_i are the linear velocity and angular velocity of the ith follower.

According to Eq. (5.1), the longitudinal and lateral velocities $\dot{x}_{il}$ and $\dot{y}_{il}$ can be transformed into the corresponding linear velocity and angular velocity through the following transformation

$$\begin{aligned} v_i &= \dot{x}_{il} \cos(\theta_i) + \dot{y}_{il} \sin(\theta_i), \\ w_i &= \left(-\dot{x}_{il} \sin(\theta_i) + \dot{y}_{il} \cos(\theta_i)\right) / l_i. \end{aligned} \tag{5.2}$$

Combining the velocity distribution [5], it is reasonable to consider the longitudinal and lateral velocities $v_{ix} \triangleq v_i \cos(\theta_i) - l_i w_i \sin(\theta_i)$ and $v_{iy} \triangleq v_i \sin(\theta_i) + l_i w_i \cos(\theta_i)$ as the design control inputs to ensure safe coordination. Referring to Eq. (5.2), v_{ix} and v_{iy} can be converted into the corresponding linear velocity and angular velocity v_i and w_i.

The modular unit leader adopts the same wheel actuation modality, and the motion control model is specifically as follows

$$\begin{cases} \dot{x}_{0l} &= v_0 \cos(\theta_0) - l_0 w_0 \sin(\theta_i), \\ \dot{y}_{0l} &= v_i \sin(\theta_i) + l_0 w_0 \cos(\theta_0), \\ \dot{\theta}_{0l} &= w_0, \end{cases} \tag{5.3}$$

The longitudinal and lateral velocities of the leader $v_{0x} \triangleq v_0 \cos(\theta_0) - l_0 w_0 \sin(\theta_0)$ and $v_{0y} \triangleq v_0 \sin(\theta_0) + l_0 w_0 \cos(\theta_0)$ are used as the design control inputs to ensure safe tracking of the reference trajectory.

This chapter can reconstruct the swarm modular unit, which is also subject to the velocity and acceleration nested saturation constraints of Eq. (4.4). Similarly, the formation configuration is determined by the orthogonal coordinates (p_{ix}, p_{iy}), using the distributed topology and control structure based on the leader-following described in Sect. 3.1, requiring that the modular units in the coordination smoothly avoid nearby obstacles and adjacent coordination modular unit formation members while performing the tasks of leader-following and formation maintenance.

Remark 5.1 In Sect. 3.2, a coordinated control of the modular unit is achieved using a kinematic and dynamic joint control framework. The dynamic controller of the modular unit in the formation can accurately track the kinematic velocity command constrained by saturation, achieving high-precision wheel velocity control. Therefore, in scenarios involving only kinematic control, the constrained velocity after saturation can be regarded as the actual velocity of the modular unit in the formation that can be reconstructed. This section still uses the dynamic controller in Sect. 3.2 at the dynamic level, and designs a safe formation control strategy under nested saturation constraints at the kinematic level under the same communication network as in Sect. 3.2. To simplify the design and analysis process of the controller, under the high-precision velocity tracking of the bottom dynamic controller, the kinematic level velocity command saturated in practical engineering can be regarded as the actual velocity of the modular unit.

Lemma 5.1 *Consider the following system and corresponding sets* $\mathcal{X}$, $\mathcal{X}_0$ *and* $\mathcal{X}_u$

$$\dot{X} = f(X), X \in \mathcal{X} \subseteq \boldsymbol{R}^n, \tag{5.4}$$

where $f(X)$ *is sufficiently smooth, and the initial condition set* $\mathcal{X}_0 \subset \mathcal{X}$ *and unsafe set* $\mathcal{X}_u \subset \mathcal{X}$ *are given. For any given* $\alpha \in \boldsymbol{R}_{+}$, *if there exists a barrier function, i.e., a continuously differentiable function* $\mathcal{B}(X) : \mathcal{X} \to \boldsymbol{R}$, *satisfying the following conditions*

$$\begin{cases} \forall X \in \mathcal{X}_0 : \mathcal{B}(X) \le 0, \\ \forall X \in \mathcal{X}_u : \mathcal{B}(X) > 0, \\ \forall X \in \mathcal{X} : \dot{\mathcal{B}}(X) \le -\alpha \mathcal{B}(X), \dot{\mathcal{B}}(X) = \dfrac{\partial \mathcal{B}}{\partial X} f(X), \end{cases} \tag{5.5}$$

then, if all state trajectories starting from $\mathcal{X}_0$ avoid the unsafe set under all $t > 0$ conditions, safety is guaranteed [6].

Assumption 5.1 *The linear and angular velocities of the navigational units in the reconfigurable swarm satisfy the velocity and acceleration saturation values, while ensuring that the reference shape configuration p_{ix} and p_{iy} and their derivatives remain within limits.*

Assumption 5.2 *For the proposed leader-follower formation system under distributed topology, its communication topology graph $\bar{\mathcal{G}}$ contains at least one directed spanning tree (in this case $\mathcal{B} \neq 0$) [7].*

5.3 Safety Formation Control and Oscillation Suppression Under Environmental Constraints

The distributed control framework based on the leader-follower structure is shown in Fig. 5.1, ensuring the safety of the reconfigurable swarm formation, by resolving collision threats from obstacles and nearby vehicles as well as severe oscillations caused by nested input saturation in dynamics. A hybrid safety strategy and adaptive anti-saturation method are used to achieve smoothness and safety of the formation.

5.3.1 Hybrid Reactive Safety Mechanism Design

(1) Safety Protection Strategy Based on Improved Artificial Potential Field

A hybrid safety cooperative strategy based on safety barriers and improved Artificial Potential Field (APF) is used to generate cooperative actions, effectively avoiding nearby obstacles or adjacent formation members. The low-trigger safety avoidance mechanism based on potential is shown in Fig. 5.2.

Drawing on the collision risk prediction mechanism observed in [8], the obstacle and collision avoidance strategy for the ith follower based on APF is designed as follows to optimize its avoidance process:

$$\begin{cases} \theta_{iko} < \varphi_{iko}, \\ r_{ko} \leq p_{iko} \leq R_{ko}, \\ |v_i| \cos \theta_{iko} \geq 0. \end{cases} \tag{5.6}$$

$$\begin{cases} \theta_{ij} < \varphi_{ij}, \\ r_{ij} \leq p_{ij} \leq R_{ij}, \\ |v_i| \cos \theta_{ij} + |v_j| \cos \theta_{ji} \geq 0. \end{cases} \tag{5.7}$$

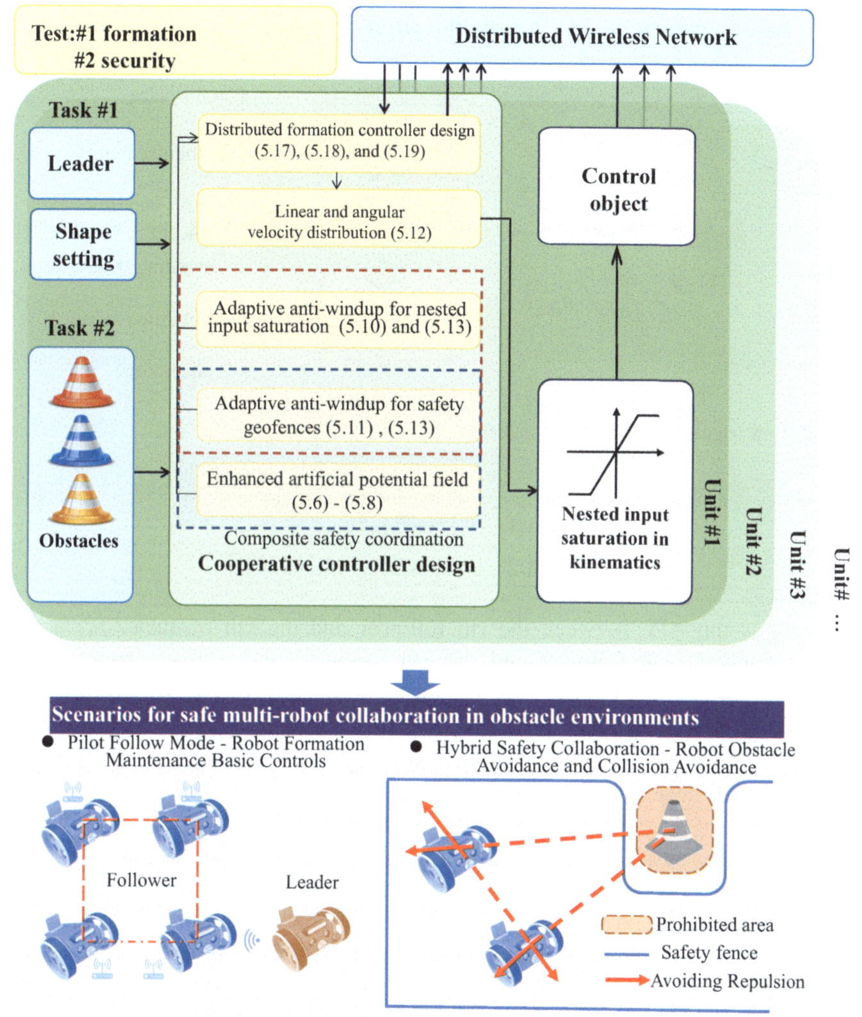

Fig. 5.1 Security cooperative control framework

$$p_{ik}^{o} = \begin{cases} -\dfrac{\alpha_{iko}\varrho_{ik}^{o}\left(p_{iko}^{2}-R_{ko}^{2}\right)^{2}}{2\left(p_{iko}^{2}-r_{ko}^{2}\right)^{2}}, & (5.6) \\ 0, & \text{otherwise} \end{cases} \tag{5.8}$$

$$p_{ij}^{v} = \begin{cases} -\dfrac{\alpha_{ijv}\varrho_{ij}\left(p_{ij}^{2}-R_{ij}^{2}\right)^{2}}{2\left(p_{ij}^{2}-r_{ij}^{2}\right)^{2}}, & (5.7) \\ 0, & \text{otherwise} \end{cases} \tag{5.9}$$

In the above equations,

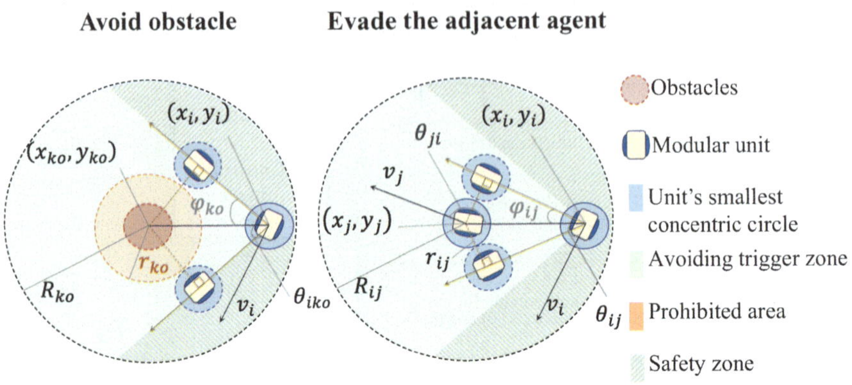

Fig. 5.2 Low trigger safe escape mechanism

$$\varrho^o_{ik} = 1 - \exp\left(-|v_i|\cos\theta_{iko}\right),$$
$$\varrho_{ij} = 1 - \exp\left(-|v_i|\cos\theta_{ij} - |v_j|\cos\theta_{ji}\right),$$

Here, p^o_{ik} is the APF between the ith follower and the kth obstacle, p^v_{ij} is the APF between the ith follower and the jth follower; $p_{iko} = \|p_i - p_{ko}\|$, where $p_i = [x_{il}, y_{il}]^{\mathrm{T}}$ and $p_{ko} = [x_{ko}, y_{ko}]^{\mathrm{T}}$ are the distances between the ith follower and the kth obstacle, x_{ko} and y_{ko} are the positions of the kth obstacle; $p_{ij} = \|p_i - p_j\|$, where $p_i = [x_{il}, y_{il}]^{\mathrm{T}}$ and $p_j = \left[x_{jl}, y_{jl}\right]^{\mathrm{T}}$ are the distances between the ith follower and the jth follower; θ_{iko} represents the LOS angle of the kth obstacle in the coordinate system of the ith follower, θ_{ij} represents the LOS angle of the jth follower in the coordinate system of the ith follower; if θ_{iko} is outside the safe angle φ_{ko}, or θ_{ij} is outside the safe angle φ_{ij}, then the ith follower has no collision risk with the adjacent obstacles or units; R_{ko} and R_{ij} are the distances from the ith follower to the kth obstacle and the jth follower that trigger collision avoidance, r_{ko} and r_{ij} are the minimum safe distances from the ith follower to the kth obstacle and the jth follower. The parameters $\alpha_{iko}, \alpha_{ijv} \in \mathbf{R}_+$ are used to adjust the response strength of the APF, and will be set to appropriate values to weaken the response action of the reconfigurable swarm units during the approach process, this improvement aims to enhance the smoothness of the unit formation control.

The overall APF of the ith follower can be written in the following form

$$p^x_i = \sum_{j=1}^{m} a_{ij} o^x_{ij} p^v_{ij} + \sum_{k=1}^{N_o} o^x_{iko} p^o_{ik}, \tag{5.10}$$

$$p^y_i = \sum_{j=1}^{m} a_{ij} o^y_{ij} p^v_{ij} + \sum_{k=1}^{N_o} o^y_{iko} p^o_{ik}, \tag{5.11}$$

where N_o represents the total number of nearby obstacles, and

$$o_{ij}^x = \operatorname{sgn}\left(\frac{\partial p_{ij}^v}{\partial x_{il}}\right), o_{ij}^y = \operatorname{sgn}\left(\frac{\partial p_{ij}^v}{\partial y_{il}}\right),$$
$$o_{iko}^x = \operatorname{sgn}\left(\frac{\partial p_{ik}^o}{\partial x_{il}}\right), o_{iko}^y = \operatorname{sgn}\left(\frac{\partial p_{ik}^o}{\partial y_{il}}\right),$$

represent the directions of the total APF in the vertical and lateral positions respectively.

Remark 5.2 The coordinated actions and formation maintenance based on the improved APF can effectively implement the mutual collision between nearby followers (5.7). Moreover, compared with the non-adaptive APF based on collision avoidance distance trigger [9], the improvement of the proposed APF lies in significantly reducing the number of triggers, allowing flexible adjustment of the APF strength, and directly applying to the kinematic model to correct the formation trajectory and control input. However, for obstacles as non-cooperative targets, relying solely on the improved APF for avoidance still easily triggers condition (5.6), and the potential of APF increases rapidly at close distances, which can easily lead to severe saturation oscillations. Therefore, inspired by the safety barrier proposed by Ghaffari [6], a hybrid safety strategy based on the acceleration envelope generated by the safety barrier and the low-energy APF triggered at close distances is designed to improve the smooth safety of the unit formation. In addition, followers with directional communication with the leader can also introduce the leader's position information into formulas (5.6), (5.7), (5.8) and (5.9) to generate behavior to avoid the leader. Since the leader does not communicate with other units, it only needs to ensure the safety of obstacle avoidance and mark the potential as p_0^x and p_0^y.

(2) Composite Obstacle Avoidance Strategy Based on Dynamic Safety Barrier
Assume that the number of obstacles in the task environment is finite [10]. At the same time, based on the position of the kth obstacle, the upper and lower safety boundaries x_{iko}^u, x_{iko}^l, y_{iko}^u, and y_{iko}^l for the ith follower to move safely in the longitudinal and lateral directions can be designed. Moreover, the safety boundaries and their first and second derivatives are all bounded. Here, considering the nested saturation of the modular unit's velocity and acceleration, we have $|\dot{x}_{iko}^u|$, $|\dot{x}_{iko}^l|$, $|\dot{y}_{iko}^u|$ and $|\dot{y}_{iko}^l| < \bar{\nu}_i \in \mathbf{R}_+$, $|\ddot{x}_{iko}^u|$, $|\ddot{x}_{iko}^l|$, $|\ddot{y}_{iko}^u|$ and $|\ddot{y}_{iko}^l| < \bar{\alpha}_i \in \mathbf{R}_+$, and $\bar{\nu}_i \leq \bar{v}_i$ and $\bar{\alpha}_i \leq \bar{a}_i$.

The ith modular unit follower designs the obstacle function series for the upper and lower bounds of the kth obstacle in the longitudinal direction as follows

$$\mathcal{F}_{iko}^{ux} = \dot{\bar{x}}_{iko}^u - h\left(-\bar{x}_{iko}^u\right) g_i\left(\bar{x}_{iko}^u\right) + m_i\left(\bar{x}_{iko}^u\right) \bar{x}_{iko}^u, \tag{5.12}$$
$$\mathcal{F}_{iko}^{lx} = \dot{\bar{x}}_{iko}^l - h\left(-\bar{x}_{iko}^l\right) g_i\left(\bar{x}_{iko}^l\right) + m_i\left(\bar{x}_{iko}^l\right) \bar{x}_{iko}^l, \tag{5.13}$$

and

$$h(x)=\begin{cases}1, & x\geq 0\\ 0, & x<0\end{cases},$$

$$m_i(x)=(1-h(-x))\sqrt{\frac{\gamma_i}{\rho_i}},$$

$$g_i(x)=\left(2\gamma_i\rho_i\ln\left(\cosh\left(\frac{x}{\rho_i}\right)\right)\right)^{\frac{1}{2}}, x\leq 0,$$

where $\bar{x}^u_{iko}=x_{il}-x^u_{iko}$ and $\bar{x}^l_{iko}=x^l_{iko}-x_{il}$; $\gamma_i,\rho_i\in\mathbf{R}_+$ are design gains, and $\gamma_i\leq\bar{\alpha}_i$.

To ensure that $\bar{x}^u_{iko}:\Omega^u_x=\{\mathcal{F}^{ux}_{iko}\leq 0\cap\bar{x}^u_{iko}\leq 0\}$ and $\bar{x}^l_{iko}:\Omega^l_x=\{\mathcal{F}^{lx}_{iko}\leq 0\cap\bar{x}^l_{iko}\leq 0\}$ are forward invariant sets, the safety boundaries of the rate of change of velocity command, i.e., the safety acceleration envelopes a^{tx}_{iko} and a^{ty}_{iko}, $t=u,l$, are designed as follows.

The upper and lower safety acceleration envelopes a^{ux}_{iko} and a^{lx}_{iko} in the longitudinal direction for the ith modular unit follower are designed as follows

$$a^{ux}_{iko}=-\alpha_i\mathcal{F}^{ux}_{iko}+\Psi^{ux}_{iko}\dot{\bar{x}}^u_{iko}+\ddot{x}^u_{iko}, \tag{5.14}$$

$$a^{lx}_{iko}=-\alpha_i\mathcal{F}^{lx}_{iko}+\Psi^{lx}_{iko}\dot{\bar{x}}^l_{iko}-\ddot{x}^l_{iko}, \tag{5.15}$$

where

$$\Psi^{ux}_{iko}=h\left(-\bar{x}^u_{iko}\right)\gamma_i\tanh\left(\frac{\bar{x}^u_{iko}}{\rho_i}\right)/g_i\left(\bar{x}^u_{iko}\right)-m_i\left(\bar{x}^u_{iko}\right),$$

$$\Psi^{lx}_{iko}=h\left(-\bar{x}^l_{iko}\right)\gamma_i\tanh\left(\frac{\bar{x}^l_{iko}}{\rho_i}\right)/g_i\left(\bar{x}^l_{iko}\right)-m_i\left(\bar{x}^l_{iko}\right),$$

where $\alpha_i\in\mathbf{R}_+$ is a design parameter.

Following the control input constraint proposed by Ghaffari [6], the acceleration of the formation unit is subjected to the saturation constraint $\operatorname{sat}\left(-a^{lx}_{iko},\ddot{x}_{il},a^{ux}_{iko}\right)$, which can ensure its longitudinal safety.

Similarly, the lateral safety accelerations a^{uy}_{iko} and a^{ly}_{iko} have the same safety protection effect. At the same time, it is necessary to apply longitudinal and lateral saturation to the modular unit, represented as $\operatorname{sat}\left(-\bar{\alpha}_i,\operatorname{sat}\left(-a^{lx}_{iko},\ddot{x}_{il},a^{ux}_{iko}\right),\bar{\alpha}_i\right)$ and $\operatorname{sat}\left(-\bar{\alpha}_i,\operatorname{sat}\left(-a^{ly}_{iko},\ddot{y}_{il},a^{uy}_{iko}\right),\bar{\alpha}_i\right)$.

The safety of the formation unit follower under unsaturated conditions is analyzed using the longitudinal upper bound as an example. If $\ddot{x}_{il}\leq a^{ux}_{iko}$, then $\dot{\mathcal{F}}^{ux}_{iko}\leq-\alpha_i\mathcal{F}^{ux}_{iko}$. In practice, it is easy to set the initial condition $\bar{x}^u_{iko}(0)\in\Omega^u_x$ so that $\mathcal{F}^{ux}_{iko}(0)\leq 0$. Therefore, according to Lemma 5.1, for all $t>0$, $\mathcal{F}^{ux}_{iko}(0)\leq 0$. In addition, when $\bar{x}^u_{iko}\to 0$, $\dot{\bar{x}}^u_{iko}+\bar{x}^u_{iko}\sqrt{\gamma_i/\rho_i}\leq 0$ and from Eq. (5.14), $\dot{\bar{x}}^u_{iko}\leq-\sqrt{\gamma_i/\rho_i}\bar{x}^u_{iko}$, which shows that for all $t\geq 0$, $\bar{x}^u_{iko}\leq 0$.

Combining the above analysis, under the constraint of the safety acceleration envelope on $\ddot{x}_{il}$, it can be ensured that $x_{il} < x^u_{iko}$, that is, the set safety upper limit will not be reached. Even if an additional disturbance temporarily causes $\bar{x}^u_{iko} \geq 0$, it will still converge to $\bar{x}^u_{iko} = 0$. In practical scenarios, the improved APF can avoid crossing the safety boundary and colliding with obstacles by adjusting the repulsive potential within a close distance. Even if additional disturbances temporarily cause $\bar{x}^u_{iko} \geq 0$, it will quickly converge to $\bar{x}^u_{iko} = 0$.

The safety of the formation follower under saturation conditions is analyzed using the vertical upper bound as an example. First, consider the case where $a^{ux}_{iko} \geq \bar{\alpha}_i$ and $\ddot{x}_{il} \geq \bar{\alpha}_i$. Then, Eq. (5.14) gives $S^{ux}_{iko} \leq 0$, where

$$S^{ux}_{iko} = \bar{\alpha}_i + \alpha_i \mathcal{F}^{ux}_{iko} - \gamma_i \tanh\left(\frac{\bar{x}^u_{iko}}{\rho_i}\right) \dot{\bar{x}}^u_{iko} / g_i\left(\bar{x}^u_{iko}\right) - \ddot{x}^u_{iko}.$$

Because within the safety set $\alpha_i \mathcal{F}^{ux}_{iko} \leq 0$, we have $\dot{\bar{x}}^u_{iko}/g_i(\bar{x}^u_{iko}) \leq 1$ and $-1 \leq \tanh\left(\bar{x}^u_{iko}/\rho_i\right) \leq 0$. Therefore, $S^{ux}_{iko} \leq 0$ leads to $\mathcal{F}^{ux}_{iko} \leq -\left(\bar{\alpha}_i + \gamma_i - \ddot{x}^u_{iko}\right)/\alpha_i$. The state starting from $S^{ux}_{iko} \leq 0$ will move between $S^{ux}_{iko} = 0$ and $\mathcal{F}^{ux}_{iko} = 0$, where the control input equals a^{ux}_{iko}. Therefore, we get $\dot{\mathcal{F}}^{tu}_{iko} = -\alpha_i \mathcal{F}^{tu}_{iko}$, which means the state has reached the barrier function, i.e., $\mathcal{F}^{tu}_{iko} = 0$. Therefore, the safety of the system can be ensured. Next, consider the case where $\ddot{x}_{il} < -\bar{\alpha}_i$ and $\ddot{x}^u_{iko} + \bar{\alpha}_i - \gamma_i > 0$, we can have $a^{ux}_{iko} > -\bar{\alpha}_i$. Finally, consider the case where $a^{ux}_{iko} < -\bar{\alpha}_i$, $\ddot{x}_{il} < -\bar{\alpha}_i$ and $\ddot{x}^u_{iko} + \bar{\alpha}_i - \gamma_i < 0$. Then, Eq. (5.14) gives $\bar{S}^{ux}_{iko} \leq 0$, where

$$\bar{S}^{ux}_{iko} = \bar{\alpha}_i - \alpha_i \mathcal{F}^{ux}_{iko} + \gamma_i \tanh\left(\frac{\bar{x}^u_{iko}}{\rho_i}\right) \dot{\bar{x}}^u_{iko} / g_i\left(\bar{x}^u_{iko}\right) + \ddot{x}^u_{iko}.$$

Therefore, $\bar{S}^{ux}_{iko} \leq 0$ leads to $\mathcal{F}^{ux}_{iko} \geq \left(\bar{\alpha}_i - \gamma_i + \ddot{x}^u_{iko}\right)/\alpha_i$. Similarly, the state starting from $\bar{S}^{ux}_{iko} \leq 0$ will move between $\bar{S}^{ux}_{iko} = 0$ and $\mathcal{F}^{ux}_{iko} = 0$, where the control input equals a^{ux}_{iko}, and we get $\dot{\mathcal{F}}^{ux}_{iko} = -\alpha_i \mathcal{F}^{ux}_{iko}$, which also means the state has reached the barrier function, i.e., $\mathcal{F}^{ux}_{iko} = 0$. Therefore, the safety of the system under acceleration saturation conditions can be effectively ensured. When applying the safety acceleration envelope, it is necessary to consider the nested saturation of velocity and acceleration at the same time. Combined with the nested saturation model [6, 11], the new saturation command constraint design is as follows:

$$\begin{aligned} \dot{v}^x_{is} &= \mathrm{sat}\left(-\bar{\alpha}_i, \mathrm{sat}\left(-\alpha^{lx}_i, \left(v^x_{is} - v_{ix}\right)/\tau_s, \alpha^{ux}_i\right), \bar{\alpha}_i\right), \\ \dot{v}^y_{is} &= \mathrm{sat}\left(-\bar{\alpha}_i, \mathrm{sat}\left(-\alpha^{ly}_i, \left(v^y_{is} - v_{iy}\right)/\tau_s, \alpha^{uy}_i\right), \bar{\alpha}_i\right), \end{aligned} \tag{5.16}$$

and

$$\begin{aligned} v^x_{is} &= \mathrm{sat}\left(-\bar{v}_i, v_{ix}, \bar{v}_i\right), \\ v^y_{is} &= \mathrm{sat}\left(-\bar{v}_i, v_{iy}, \bar{v}_i\right), \end{aligned}$$

where v_{ix} and v_{iy} are the longitudinally and laterally designed velocity commands; τ_s is the system time constant; v_{is}^x and v_{is}^y are the safe saturation velocities that satisfy the nested saturation of velocity and acceleration in kinematics, used to constrain the velocity not to exceed the new nested saturation, and to stimulate the modular unit to avoid adjacent obstacles.

From Eqs. (5.1) and (5.16), we can get

$$\begin{aligned} v_i &= v_{is}^x \cos(\theta_i) + v_{is}^y \sin(\theta_i), \\ w_i &= \left(v_{is}^y \cos(\theta_i) - v_{is}^x \sin(\theta_i)\right)/l_i, \end{aligned} \tag{5.17}$$

and both v_i and w_i are limited by the amplitude and rate nested saturation constraints in (4.4).

In Eq. (5.16), the safety acceleration envelope with nested velocity and acceleration saturation differs from the previous safety acceleration envelope with only acceleration saturation in two main aspects. First, under the premise of system stability, due to the influence of uncertainty and nested linear/angular velocity saturation, $\dot{v}_{ix}^s$ and $\dot{v}_{iy}^s$ exhibit bounded and adjustable deviations. Secondly, due to the above factors, v_{ix}^s and v_{iy}^s also show bounded and adjustable deviations. These two layers of deviation can be considered as external disturbances in the system in Lemma 5.1. According to the convergence analysis of $\dot{\mathcal{F}}_{iko}^{ux} \leq -\alpha_i \mathcal{F}_{iko}^{ux}$, even in the presence of disturbances, the state will still converge to the safety barrier. When the improved APF and safety barrier design are applied together, this method effectively prevents the unit from colliding with obstacles. At the same time, due to the safety envelope's constraint on the control input, it is easy to trigger saturation-induced oscillations. Therefore, it is necessary to design an adaptive auxiliary dynamic system to suppress oscillations and smooth the formation trajectory. In addition, the leader unit can also use the nested saturation in (5.16) and the improved APF to drive it to avoid nearby obstacles, marking the safety constraints as v_{0s}^x and v_{0s}^y.

5.3.2 *Adaptive Oscillation Suppression Distributed Formation Controller*

Combining nested input saturation and safety acceleration constraints, the adaptive auxiliary dynamic system design for the leader and the ith ($i = 1, \ldots, m$) unit follower is as follows

$$\begin{cases} \dot{\lambda}_{0x} = \cosh^2(\lambda_{0x})\left(-k_{0x}\lambda_{0x} + \Delta_{0x}\right)/\gamma_0, \\ \dot{\lambda}_{iy} = \cosh^2\left(\lambda_{iy}\right)\left(-k_{0y}\lambda_{0y} + \Delta_{0y}\right)/\gamma_0. \end{cases} \tag{5.18}$$

$$\begin{cases} \dot{\lambda}_{0s}^x = \cosh^2\left(\lambda_{0s}^x\right)\left(-k_{0s}^x\lambda_{0s}^x + \Delta_{0s}^x\right)/\gamma_{0s}, \\ \dot{\lambda}_{0s}^y = \cosh^2\left(\lambda_{0s}^y\right)\left(-k_{0s}^y\lambda_{0s}^y + \Delta_{0s}^y\right)/\gamma_{0s}. \end{cases} \tag{5.19}$$

$$\begin{cases} \dot{\lambda}_{ix} = \cosh^2(\lambda_{ix})\left(-k_{ix}\lambda_{ix} + \Delta_{ix}\right)/\gamma_i, \\ \dot{\lambda}_{iy} = \cosh^2\left(\lambda_{iy}\right)\left(-k_{iy}\lambda_{iy} + \Delta_{iy}\right)/\gamma_i. \end{cases} \tag{5.20}$$

$$\begin{cases} \dot{\lambda}_{is}^{x} = \cosh^2\left(\lambda_{is}^{x}\right)\left(-k_{is}^{x}\lambda_{is}^{x} + \Delta_{is}^{x}\right)/\gamma_{is}, \\ \dot{\lambda}_{is}^{y} = \cosh^2\left(\lambda_{is}^{y}\right)\left(-k_{is}^{y}\lambda_{is}^{y} + \Delta_{is}^{y}\right)/\gamma_{is}. \end{cases} \tag{5.21}$$

And

$$\begin{aligned} \Delta_{0x} &= (v_{0s} - v_0)\cos(\theta_0) - l_i(w_{0s} - w_0)\sin(\theta_0), \\ \Delta_{0y} &= (v_{0s} - v_0)\sin(\theta_0) + l_i(w_{0s} - w_0)\cos(\theta_0), \\ \Delta_{0s}^{x} &= v_{0s}^{x} - v_{0x}, \Delta_{0s}^{y} = v_{0s}^{y} - v_{0y}, \\ \Delta_{ix} &= (v_{is} - v_i)\cos(\theta_i) - l_i(w_{is} - w_i)\sin(\theta_i), \\ \Delta_{iy} &= (v_{is} - v_i)\sin(\theta_i) + l_i(w_{is} - w_i)\cos(\theta_i), \\ \Delta_{is}^{x} &= v_{is}^{x} - v_{ix}, \Delta_{is}^{y} = v_{is}^{y} - v_{iy}, \end{aligned}$$

where λ_{0x}, λ_{0y}, λ_{0x}^{s} and λ_{0y}^{s} are the auxiliary dynamic variables of the leader unit, $k_{0x}, k_{0y}, k_{0s}^{x}, k_{0s}^{y}, \gamma_i$ and $\gamma_{0s} \in \mathbf{R}_+$ are the corresponding adaptive gains; λ_{ix}, λ_{iy}, λ_{ix}^{s} and λ_{iy}^{s} are the auxiliary dynamic variables of the follower unit, $k_{ix}, k_{iy}, k_{is}^{x}, k_{is}^{y}, \gamma_i$ and $\gamma_{is} \in \mathbf{R}_+$ are the corresponding adaptive gains.

The formation trajectory errors z_{ix} and z_{iy} of the leader and the ith follower $(i = 1, \ldots, m)$ are defined as follows

$$\begin{aligned} z_{0x} &= x_{0l} - x_r - \gamma_0 \tanh(\lambda_{0x}) - \gamma_{0s}\tanh(\lambda_{0s}^{x}) + p_0^{x}, \\ z_{0y} &= y_{0l} - y_r - \gamma_0 \tanh(\lambda_{0y}) - \gamma_{0s}\tanh(\lambda_{0s}^{y}) + p_0^{y}. \end{aligned} \tag{5.22}$$

$$\begin{aligned} z_{ix} &= x_{il} - x_{0l} - p_{ix} - \gamma_i \tanh(\lambda_{ix}) - \gamma_{is}\tanh(\lambda_{is}^{x}) + p_i^{x}, \\ z_{iy} &= y_{il} - y_{0l} - p_{iy} - \gamma_i \tanh(\lambda_{iy}) - \gamma_{is}\tanh(\lambda_{is}^{y}) + p_i^{y}. \end{aligned} \tag{5.23}$$

Therefore, the longitudinal and lateral position coordination errors e_{1i} and e_{2i} are written in the following form

$$\begin{aligned} e_{1i} &= b_i z_{ix} + \sum_{j=1}^{m} a_{ij}\left(z_{ix} - z_{jx}\right), \\ e_{2i} &= b_i z_{iy} + \sum_{j=1}^{m} a_{ij}\left(z_{iy} - z_{jy}\right). \end{aligned} \tag{5.24}$$

The design of longitudinal and lateral velocity commands for leader trajectory tracking, follower navigation tracking, formation maintenance, and coordinated obstacle avoidance are as follows:

$$\begin{aligned} v_{0x} &= -k_{10}z_{0x} + \dot{x}_r - \dot{p}_0^{x} - k_{0s}^{x}\lambda_{0s}^{x} - k_{0x}\lambda_{0x}, \\ v_{0y} &= -k_{20}z_{0y} + \dot{y}_r - \dot{p}_0^{y} - k_{0s}^{y}\lambda_{0s}^{y} - k_{0y}\lambda_{0y}. \end{aligned} \tag{5.25}$$

$$
\begin{aligned}
v_{ix} &= -k_{1i}e_{1i} + \hat{h}_{ix} + \dot{p}_{ix} - \dot{p}_i^x - k_{is}^x \lambda_{is}^x - k_{ix}\lambda_{ix}, \\
v_{iy} &= -k_{2i}e_{2i} + \hat{h}_{iy} + \dot{p}_{iy} - \dot{p}_i^y - k_{is}^y \lambda_{is}^y - k_{iy}\lambda_{iy},
\end{aligned}
\tag{5.26}
$$

where k_{10} and $k_{20} \in \mathbf{R}_+$ are the design parameters for leader trajectory control, k_{1i} and $k_{2i} \in \mathbf{R}_+$ are the design parameters for followers, and $\hat{h}_{ix}$ and $\hat{h}_{iy}$ are the estimates of the leader's longitudinal and lateral velocities $\dot{x}_{0l}$ and $\dot{y}_{0l}$, respectively. The corresponding distributed observer design is as follows:

$$
\begin{aligned}
\dot{\hat{h}}_{ix} &= -k_{ih}^x \left(\sum_{j=1}^m a_{ij} \left(\hat{h}_{ix} - \hat{h}_{jx} \right) + b_i \left(\hat{h}_{ix} - \dot{x}_{0l} \right) \right), \\
\dot{\hat{h}}_{iy} &= -k_{ih}^y \left(\sum_{j=1}^m a_{ij} \left(\hat{h}_{iy} - \hat{h}_{jy} \right) + b_i \left(\hat{h}_{iy} - \dot{y}_{0l} \right) \right),
\end{aligned}
\tag{5.27}
$$

where k_{ih}^x and k_{ih}^y are the positive observer gains.

5.3.3 Stability Analysis

Theorem 5.1 *Consider the single-agent swarm system based on the leader-follower structure (5.1), (5.3), subject to the constraints of nested input saturation of velocity and acceleration (4.4) and safe acceleration envelope (5.16). Assuming assumptions 5.1 and 5.2 hold, then the navigation tracking controller (5.25) and the distributed formation controller (5.26)–(5.27), as well as the APF (5.6)–(5.9) used to avoid neighboring obstacles and adjacent agents, and the adaptive anti-saturation system (4.24) used to suppress saturation oscillations, can ensure that the leader tracking error e_{0z} asymptotically converges to 0, and the formation tracking error e_z of the followers eventually converges to the invariant set region*

$$
\Omega_z = \{e_z \in \boldsymbol{R}^{2m} | V_z \le \frac{4\bar{\delta}_z^2 \| P_z \| \lambda_{\max}^2 (P_z)}{\lambda_{\min}^2 (Q_z)}\},
$$

Here, $e_z = col\{z_x, z_y\}$, $z_x = [z_{1x}, \ldots, z_{mx}]^T$ and $z_y = \left[z_{1y}, \ldots, z_{my}\right]^T$; $K_1 = diag\{k_{11}, \ldots, k_{1m}\}$ and $K_2 = diag\{k_{21}, \ldots, k_{2m}\}$ are positive definite gain matrices; the symmetric positive definite matrix related to the communication topology of the modular unit swarm is $P_z = diag\{0.5\mathcal{M}, 0.5\mathcal{M}\}$; $Q_z = diag\{\mathcal{M}K_1\mathcal{M}, \mathcal{M}K_2\mathcal{M}\}$ is a symmetric positive definite matrix; $\bar{\delta}_z$ is the upper bound of the leader's estimation error.

Proof The Lyapunov function of the leader-follower is chosen as follows

$$
V_{0z} = \frac{1}{2} z_{0x}^2 + \frac{1}{2} z_{0y}^2. \tag{5.28}
$$

The derivatives of the leader x_{0l} and y_{0l} based on the nested saturation are written as follows

$$\begin{aligned}\dot{x}_{0l} &= v_{0x} + \left(v_{0s}^{x} - v_{0x}\right) + (v_{0s} - v_0)\cos(\theta_0) - l_0(w_{0s} - w_0)\sin(\theta_0)\\ &= v_{0x} + \Delta_{0s}^{x} + \Delta_{0x},\\ \dot{y}_{0l} &= v_{0y} + \left(v_{0s}^{y} - v_{0y}\right) + (v_{0s} - v_0)\sin(\theta_0) + l_0(w_{0s} - w_0)\cos(\theta_0)\\ &= v_{0y} + \Delta_{0s}^{y} + \Delta_{0y}.\end{aligned} \tag{5.29}$$

By using Eqs. (5.25) and (5.29), the derivatives of z_{0x} and z_{0y} can be obtained as follows

$$\begin{aligned}\dot{z}_{0x} &= \dot{x}_{0l} - \dot{x}_r + k_{0x}\lambda_{0x} - \Delta_{0x} + k_{0s}^{x}\lambda_{0s}^{x} - \Delta_{0s}^{x} + \dot{p}_0^{x}\\ &= v_{0x} - \dot{x}_r + k_{0x}\lambda_{0x} + k_{0s}^{x}\lambda_{0s}^{x} + \dot{p}_0^{x} = -k_{10}z_{0x},\\ \dot{z}_{0y} &= \dot{y}_{0l} - \dot{y}_r + k_{0y}\lambda_{0y} - \Delta_{0y} + k_{0s}^{y}\lambda_{0s}^{y} - \Delta_{0s}^{y} + \dot{p}_0^{y}\\ &= v_{0y} - \dot{y}_r + k_{0y}\lambda_{0y} + k_{0s}^{y}\lambda_{0s}^{y} + \dot{p}_0^{y} = -k_{10}z_{0y}.\end{aligned} \tag{5.30}$$

The derivative of V_{0z} is written as follows

$$\dot{V}_{0z} = z_{0x}\dot{z}_{0x} + z_{0y}\dot{z}_{0y} = -k_{10}z_{0x}^2 - k_{20}z_{0y}^2 \le -\bar{k}_0 V_{0z}, \tag{5.31}$$

where $\bar{k}_0 = \min\{k_{10}, k_{20}\} \in \mathbf{R}_+$. Therefore, the leader's z_{0x} and z_{0y} can theoretically converge to 0, and the convergence rate can be adjusted by controlling the gain.

The Lyapunov function of the leader's observer estimation error is defined as follows

$$V_h = \frac{1}{2}\tilde{h}_x^{\mathrm{T}}\mathcal{M}\tilde{h}_x + \frac{1}{2}\tilde{h}_y^{\mathrm{T}}\mathcal{M}\tilde{h}_y, \tag{5.32}$$

In this, $\tilde{h}_x = \left[\tilde{h}_{1x}, \ldots, \tilde{h}_{mx}\right]^{\mathrm{T}}$ and $\tilde{h}_y = \left[\tilde{h}_{1y}, \ldots, \tilde{h}_{my}\right]^{\mathrm{T}}$, $\tilde{h}_{ix} = \hat{h}_{ix} - \dot{x}_{0l}$ and $\tilde{h}_{iy} = \hat{h}_{iy} - \dot{y}_{0l}$; $e_h = \mathrm{col}\{\tilde{h}_x, \tilde{h}_y\}$; $\bar{X}_h = [\dot{x}_{0l}, \ldots, \dot{x}_{0l}]^{\mathrm{T}} \in \mathbf{R}^m$ and $\bar{Y}_h = [\dot{y}_{0l}, \ldots, \dot{y}_{0l}]^{\mathrm{T}} \in \mathbf{R}^m$. $\bar{X}_h$ and $\bar{Y}_h$ are bounded, and there exists a constant δ_h such that $\|\mathrm{col}\{\bar{X}_h, \bar{Y}_h\}\| \le \delta_h \in \mathbf{R}_+$.

According to Assumption 5.2 and Ref. [7], it is easy to conclude that $\mathcal{M}$ is symmetric positive definite.

The derivative of V_h is written in the following form

$$\begin{aligned}\dot{V}_h &= -\tilde{h}_x^{\mathrm{T}}\mathcal{M}\left(K_h^{x}\mathcal{M}\tilde{h}_x + \bar{X}_0\right) - \tilde{h}_y^{\mathrm{T}}\mathcal{M}\left(K_h^{y}\mathcal{M}\tilde{h}_y + \bar{Y}_0\right)\\ &\le -\frac{\lambda_{\min}(Q_h)}{\lambda_{\max}(P_h)}V_h + 2\delta_h\sqrt{V_h}\sqrt{P_h},\end{aligned} \tag{5.33}$$

and

$$K_h^x = \mathrm{diag}\{k_{1h}^x, \ldots, k_{mh}^x\},\ K_h^y = \mathrm{diag}\{k_{1h}^y, \ldots, k_{mh}^y\},$$
$$Q_h = \mathrm{diag}\{\mathcal{M}K_h^x\mathcal{M}, \mathcal{M}K_h^y\mathcal{M}\},\ P_h = \mathrm{diag}\{0.5\mathcal{M}, 0.5\mathcal{M}\},$$

where K_h^x, K_h^y, Q_h and P_h are positive definite matrices.

There exists a set $\Omega_h = \{e_h \in \mathbf{R}^{2m} | V_h \leq \frac{4\bar{\delta}_h^2 \| P_h \| \lambda_{\max}^2(P_h)}{\lambda_{\min}^2(Q_h)}\}$ such that for $e_h \in \{\mathbf{R}^{2m} \backslash \Omega_h\}$, $\dot{V}_h \leq 0$, thus e_h converges to Ω_h. Here, $\bar{\delta}_z$ is the upper bound of the leader's estimation error e_h.

The corresponding Lyapunov function for the follower formation coordination is defined as follows

$$V_z = \frac{1}{2} z_x^{\mathrm{T}} \mathcal{M} z_x + \frac{1}{2} z_y^{\mathrm{T}} \mathcal{M} z_y. \tag{5.34}$$

The derivatives of x_{il} and y_{il} with nested input saturation are written in the following form

$$\begin{aligned}
\dot{x}_{il} &= v_{ix} + \left(v_{is}^x - v_{ix}\right) + (v_{is} - v_i)\cos(\theta_i) - l_i(w_{is} - w_i)\sin(\theta_i) \\
&= v_{ix} + \Delta_{is}^x + \Delta_{ix}, \\
\dot{y}_{il} &= v_{iy} + \left(v_{is}^y - v_{iy}\right) + (v_{is} - v_i)\sin(\theta_i) + l_i(w_{is} - w_i)\cos(\theta_i) \\
&= v_{iy} + \Delta_{is}^y + \Delta_{iy}.
\end{aligned} \tag{5.35}$$

By using Eqs. (5.26) and (5.35), the derivatives of z_{ix} and z_{iy} can be obtained as follows

$$\begin{aligned}
\dot{z}_{ix} &= \dot{x}_{il} - \dot{x}_{0l} - \dot{p}_{ix} + k_{ix}\lambda_{ix} - \Delta_{ix} + k_{is}^x \lambda_{is}^x - \Delta_{is}^x + \dot{p}_i^x \\
&= v_{ix} - \dot{x}_{0l} - \dot{p}_{ix} + k_{ix}\lambda_{ix} + k_{is}^x \lambda_{is}^x + \dot{p}_i^x \\
&= -k_{1i}e_{1i} + \tilde{h}_{ix}, \\
\dot{z}_{iy} &= \dot{y}_{il} - \dot{y}_{0l} - \dot{p}_{iy} + k_{iy}\lambda_{iy} - \Delta_{iy} + k_{is}^y \lambda_{is}^y - \Delta_{is}^y + \dot{p}_i^y \\
&= v_{iy} - \dot{y}_{0l} - \dot{p}_{iy} + k_{iy}\lambda_{iy} + k_{is}^y \lambda_{is}^y + \dot{p}_i^y \\
&= -k_{2i}e_{2i} + \tilde{h}_{iy}.
\end{aligned} \tag{5.36}$$

The derivative of V_z is written in the following form

$$\begin{aligned}
\dot{V}_z &= z_x^{\mathrm{T}} \mathcal{M} \dot{z}_x + z_y^{\mathrm{T}} \mathcal{M} \dot{z}_y \\
&= -z_x^{\mathrm{T}} \mathcal{M} K_1 \mathcal{M} z_x - z_y^{\mathrm{T}} \mathcal{M} K_2 \mathcal{M} z_y \\
&\quad + z_x^{\mathrm{T}} \mathcal{M} \tilde{h}_x + z_y^{\mathrm{T}} \mathcal{M} \tilde{h}_y \\
&\leq -\frac{\lambda_{\min}(Q_z)}{\lambda_{\max}(P_z)} V_z + 2\bar{\delta}_z \sqrt{\| P_z \|} \sqrt{V_z} \\
&\leq -\frac{\lambda_{\min}(Q_z)}{\lambda_{\max}(P_z)} \left(\sqrt{V_z} - \frac{\lambda_{\max}(P_z) 2\bar{\delta}_z \| P_z \|}{\lambda_{\min}(Q_z)} \right) \sqrt{V_z}.
\end{aligned} \tag{5.37}$$

Obviously, when $e_z \in \{\mathbf{R}^{2m} \backslash \Omega_z\}$, $\dot{V}_z \leq 0$. Therefore, the formation tracking error e_z will eventually converge to the invariant set Ω_z, and the control precision depends on the size of the invariant set

$$\Omega_z = \{e_z \in \mathbf{R}^{2m} | V_z \leq \frac{4\bar{\delta}_z^2 \| P_z \| \lambda_{\max}^2 (P_z)}{\lambda_{\min}^2 (Q_z)}\}.$$

The size of Ω_z is determined by $\lambda_{\min}(Q_z)$. Therefore, the entire multi-agent cooperative control system is stable, and the formation tracking error is adjusted through K_h^x and K_h^y.

Remark 5.3 This chapter adopts the same control structure of kinematics and dynamics in simulation and practical application. The upper layer uses the coordinated controller designed in this chapter to generate velocity commands, and the lower layer realizes the precise tracking of velocity commands according to the dynamic controller in Chap. 4. In the cascaded control framework, the dynamic controller has a higher control bandwidth and can quickly respond to the command output of the kinematic controller. The stability of kinematic control is not significantly affected by the bottom dynamics. Therefore, it is completely feasible to design kinematic coordinated control separately using the separation principle in theory and application, which can also simplify the design process of the coordinated controller and reduce the complexity of system debugging in experiments.

5.4 Simulation and Experimental Verification

5.4.1 Simulation Result Analysis

The numerical simulation of the coordinated control of modular units was deployed on MATLAB R2021b, with a simulation duration of 100 s and a simulation step size of 1 ms. There are a total of three modular units, one of which serves as the leader, and the other two as followers in omnidirectional communication, one of which can obtain the position and velocity of the leader. The formation reference trajectory is a circular trajectory with a linear velocity of 0.4 m/s and an angular velocity of 0.1 rad/s. The leader and the followers adopt the same control architecture, but due to the formation communication topology, there is no need to maintain formation and avoid collision with nearby modular units except for tracking the reference trajectory. The size of the modular unit is set to 0.1 m, and the positions and sizes of the four circular obstacles in the simulation environment are set to

$$\begin{aligned}
(x_{1o}, y_{1o}, r_{1o}) &= (4.35, 4.0, 0.2) \text{ m}, \\
(x_{2o}, y_{2o}, r_{2o}) &= (4.00, 4.0, 0.1) \text{ m}, \\
(x_{3o}, y_{3o}, r_{3o}) &= (-3.55, 4.0, 0.1) \text{ m}, \\
(x_{4o}, y_{4o}, r_{4o}) &= (-4.38, 4.0, 0.2) \text{ m}.
\end{aligned}$$

To verify the advantages of the proposed method in reconfiguring the smoothness and safety of the swarm unit modality formation under nested saturation and environmental safety constraints, three sets of simulation experiments are designed as control groups. They respectively adopt the proposed distributed formation control method, the distributed formation control method proposed by Sharma with adaptive anti-saturation [9], and the distributed formation control method proposed by Sharma without adaptive anti-saturation [9], and are respectively labeled as Proposed, Sharma2020AW, and Sharma2020. Considering that the velocity and acceleration saturation thresholds of the formation unit are both 1.5 and 0.4, the comparison controllers use the same linear control and observer gains $k_{1i} = 0.5$, $k_{2i} = 0.5, k_{ih}^{x} = 100$ and $k_{ih}^{y} = 100$. The proposed composite nonlinear control gains are $\kappa_{1i} = \kappa_{2i} = 3$ and $\beta_{1i} = \beta_{2i} = 10$, and the adaptive gains are $k_{ix} = k_{iy} = 10$ and $k_{is}^{x} = k_{is}^{y} = 100$. The parameters related to the evasion potential function are designed as $R_{1o} = R_{4o} = 0.4$ m, $R_{2o} = R_{3o} = R_{ij} = 0.4$ m, $\alpha_{i1o} = \alpha_{i2o} = \alpha_{i3o} = \alpha_{i4o} = 0.001$ and $\alpha_{ij} = 0.1$. The design parameters related to the safety barrier are $\gamma_i = \bar{\alpha}_i = 0.4$, $\bar{v}_i = 1.0$, $\rho_i = 0.5$, $\alpha_i = 20$, $\kappa_i = 5$ and $\beta_i = 10$.

Figures 5.3 and 5.4 compare the formation plane trajectories, coordination errors, and saturation control inputs of three methods: nested saturation, obstacle avoidance, and collision avoidance under safety constraints for reconfigurable swarm unit modalities. Under the combined effects of safety barriers and avoidance potentials, the proposed method outperforms the Sharma2020AW method in terms of oscillation amplitude and recovery time before and after avoidance, exhibiting a clear smooth response characteristic. In contrast, the Sharma2020 method falls into complete oscillation under the reactive safety mechanism that is prone to trigger saturation. As can be seen from Fig. 5.4, among the saturation control inputs of the three methods, the control input of the proposed method is smoother than that of the Sharma2020AW and Sharma2020 methods, and the energy required for the safe response when the reconfigurable swarm unit modality formation faces danger is also less. The above results show that the introduction of adaptive anti-saturation can effectively suppress the saturation oscillation triggered during the avoidance process. On this basis, compared to the single repulsive avoidance potential, by combining the avoidance potential and safety barrier constraints on acceleration, the composite safety mechanism for bypassing obstacles has more advantages in maintaining formation integrity and smooth formation trajectories, making the reconfigurable swarm unit formation more flexible in responding to the threat of dangerous targets and more suitable for complex scene operation tasks.

5.4.2 *Experimental Result Analysis*

To verify the adaptability enhancement of the proposed method for the cooperative performance of the modular unit under the constraints of nested saturation and environmental safety in real scenarios, we have built a collaborative experimental

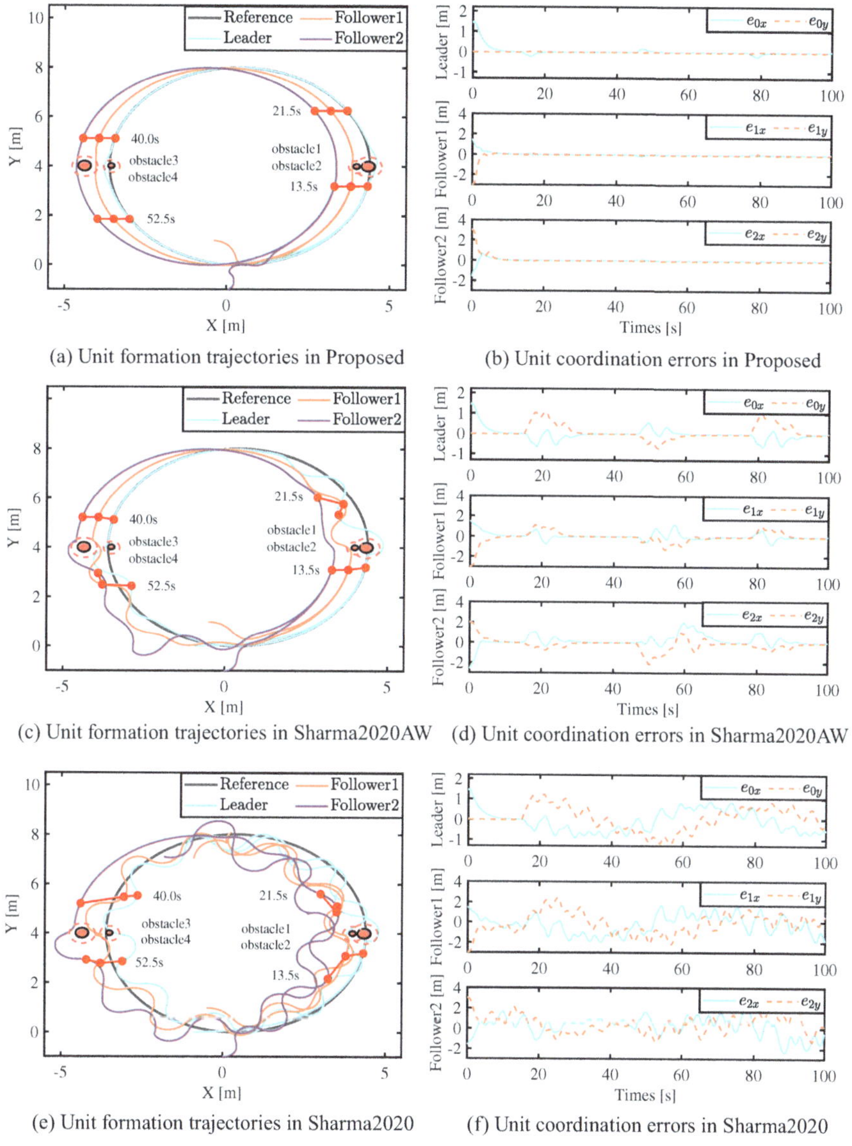

(a) Unit formation trajectories in Proposed

(b) Unit coordination errors in Proposed

(c) Unit formation trajectories in Sharma2020AW

(d) Unit coordination errors in Sharma2020AW

(e) Unit formation trajectories in Sharma2020

(f) Unit coordination errors in Sharma2020

Fig. 5.3 Comparison of safety coordination effects of multiple unit bodies

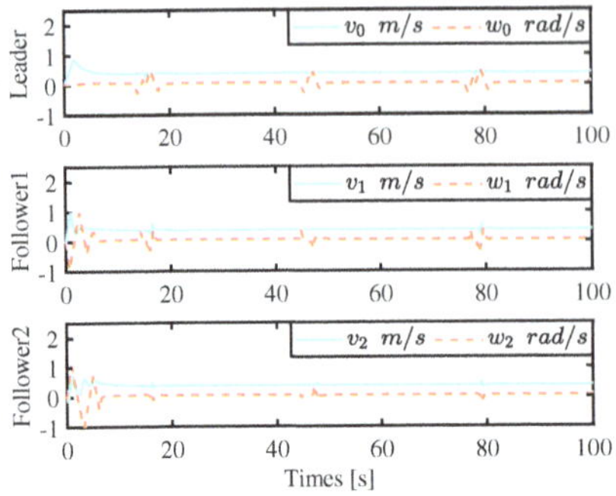

(a) Unit modality control input in the proposed method

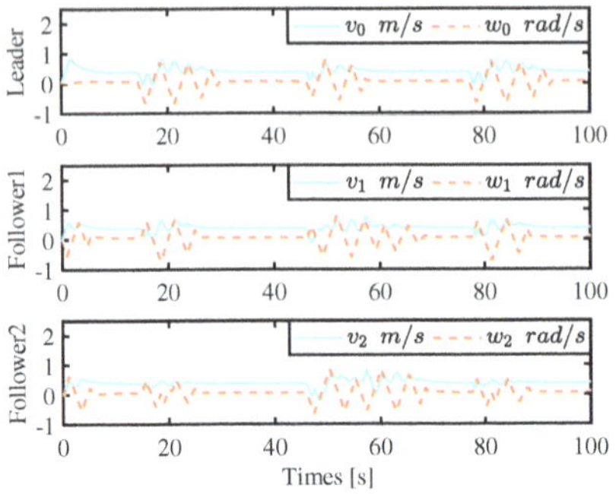

(b) Unit modality control input in Sharma2020AW method

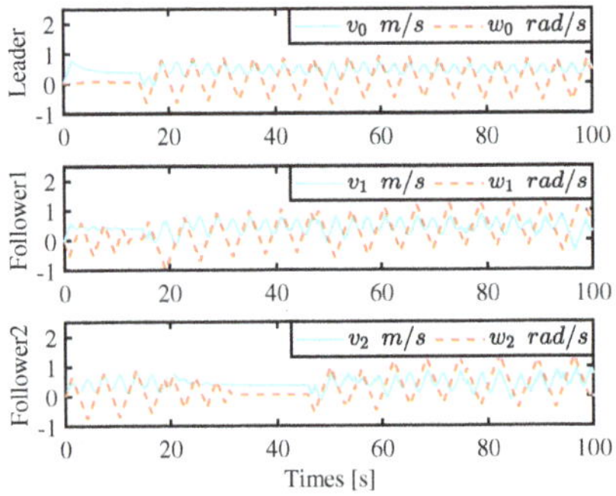

(c) Unit modality control input in Sharma2020 method

Fig. 5.4 Comparison of safety coordination kinematic control inputs of multiple unit bodies

platform(cf. Chap. 2), and carried out multi-unit formation and safety protection experiments. The formation scale is three modular units, and the communication topology is consistent with the simulation. In this section, we use a motion capture camera array to identify the Marker attached to the top of the modular unit, and calculate its motion state through image processing computers. WiFi devices and routers are used to realize the communication between each modular unit and the swarm control computer. The reference trajectory of the leader modular unit is a circular trajectory with a linear velocity of 0.08 m/s and an angular velocity of 0.05 rad/s. The follower modular units maintain an isosceles triangle formation with the leader, and the formation parameters are $\left[p_{1x}, p_{1y}, p_{2x}, p_{2y}\right] = \varrho\,[-0.4, 0.6, 0.4, 0.6]$ m. The threshold values of the velocity and acceleration of the formation modular units under nested saturation are 1.5 and 0.4, respectively. The proposed method (marked as Proposed) and the formation method proposed by Sharma [9] that integrates adaptive anti-saturation (marked as Sharma2020) are used for experimental comparison. The linear control gains, observer gains, adaptive gains, and potential function parameters of the two methods are set to be consistent. Four obstacles of different sizes and positions are placed around the field to verify and compare the safety and smoothness of the methods. The diameters and positions of the obstacles are specifically as follows

$$
\begin{aligned}
(x_{1o}, y_{1o}, r_{1o}) &= (2.53, -1.14, 0.2)\ \text{m}, \\
(x_{2o}, y_{2o}, r_{2o}) &= (4.02, -0.13, 0.2)\ \text{m}, \\
(x_{3o}, y_{3o}, r_{3o}) &= (0.52, 1.15, 0.54)\ \text{m}, \\
(x_{4o}, y_{4o}, r_{4o}) &= (-0.66, -1.54, 0.5)\ \text{m}.
\end{aligned}
$$

Figure 5.5a, c show the plane formation trajectories under nested motion saturation of two methods, where 35s, 81s, and 105s represent the moments after the modular units coordinate to avoid obstacles. The proposed method can maintain the desired triangular formation at these moments, while it is difficult for Sharma2020 to

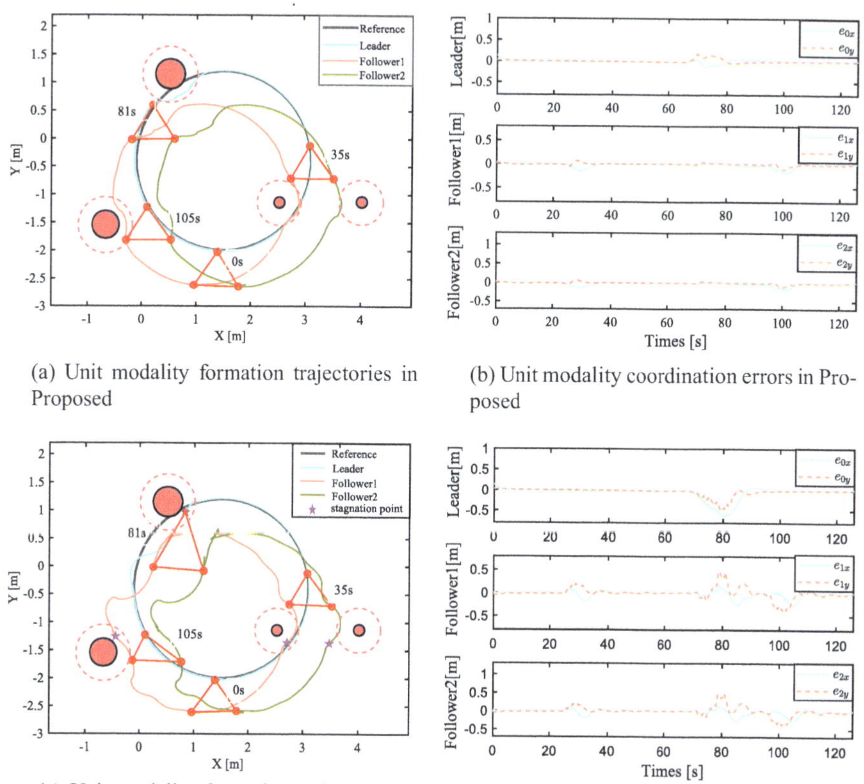

(a) Unit modality formation trajectories in Proposed

(b) Unit modality coordination errors in Proposed

(c) Unit modality formation trajectories in Sharma

(d) Unit modality coordination errors in Sharma

Fig. 5.5 Comparison of security coordination in multi-unit modes

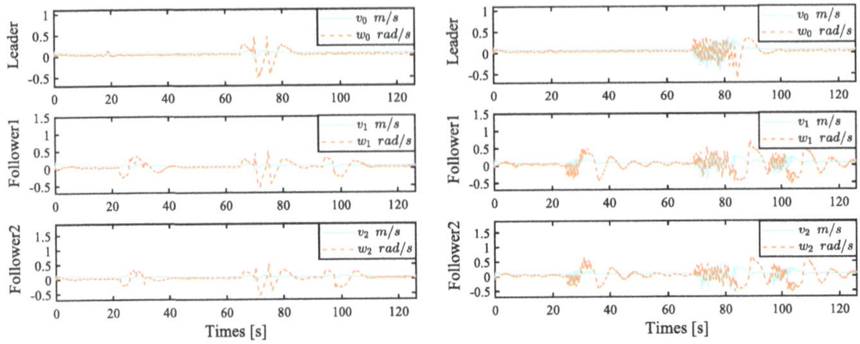

(a) Control Input of Unit modality Coordination in Proposed Method

(b) Control Input of Unit modality Coordination in Sharma Method

Fig. 5.6 Comparison of kinematic control input in multi-unit modality coordination

maintain the desired shape of the modular unit formation and has not recovered the formation performance. Compared to the severe motion stagnation and saturation oscillation of the followers in Sharma2020, the proposed method can better maintain the formation shape and recover the predetermined coordination state when avoiding obstacles. Figure 5.5b, d compare the coordination errors of the two methods. The steady-state accuracy of the leaders in both methods (≤ 4 cm) is basically the same, but when they avoid obstacles, the oscillation amplitude and recovery time in the proposed method are significantly better than Sharma2020. Figure 5.6 shows the velocity commands of the two methods. The kinematic control input of the proposed method is smoother than Sharma, and the burden on the bottom dynamic control is correspondingly less. Combined with the adaptive anti-saturation measures, compared to the single force safety mechanism of Sharma2020, the composite safety mechanism mainly bypassing obstacles has more advantages in smoothness and oscillation suppression ability on multi-unit formations, and is more suitable for complex scenes with dense obstacle distribution.

5.5 Summary

This chapter addresses the problem of distributed smoothing coordination and safety assurance of swarm units under the constraints of nested motion saturation and environmental safety in the context of reconfiguration swarm units. An adaptive enhancement method under multiple constraints of unit modalities is proposed. Firstly, a composite safety strategy of dynamic safety barriers and low-trigger potential functions is constructed to ensure the safety of unit formation members under obstacle environments. Secondly, a composite adaptive auxiliary dynamic system is introduced to smoothly suppress the severe oscillations triggered by the inherent nested motion saturation and safety acceleration envelope of the unit modality. Then, by integrating safety and anti-saturation oscillation mechanisms, a formation controller under distributed topology is designed to synchronously complete lead-following, formation maintenance, and safety assurance. Finally, a self-built reconfiguration swarm unit safety coordination simulation and experimental platform is established. Through comparative simulation and experiments, the adaptive advantages of the proposed method in dynamic response in complex environments are verified.

References

1. Wang, G., Wang, X., Li, S.: A guidance module based formation control scheme for multi-mobile robot systems with collision avoidance. IEEE Trans. Autom. Sci. Eng. **21**(1), 382–393 (2024)
2. Li, Y., Dong, S., Li, K., Tong, S.: Fuzzy adaptive fault tolerant time-varying formation control for nonholonomic multirobot systems with range constraints. IEEE Trans. Intell. Veh. **8**(6), 3668–3679 (2023)
3. Dai, S., Lu, K., Jun, F.: Fixed-time rigidity-based formation maneuvering for nonholonomic multirobot systems with prescribed performance. IEEE Trans. Cybern. **54**(4), 2129–2141 (2024)
4. Dai, S., He, S., Cai, H., Yang, C.: Adaptive leader-follower formation control of underactuated surface vehicles with guaranteed performance. IEEE Trans. Syst. Man Cybern. Syst. **52**(3), 1997–2008 (2020)
5. Chang, S., Wang, Y., Zuo, Z., Yang, H.: Fixed-time formation control for wheeled mobile robots with prescribed performance. IEEE Trans. Control Syst. Technol. **30**(2), 844–851 (2022)
6. Ghaffari, A.: Analytical design and experimental verification of geofencing control for aerial applications. IEEE/ASME Trans. Mechatron. **26**(2), 1106–1117 (2020)
7. Li, S., Wang, X.: Finite-time consensus and collision avoidance control algorithms for multiple AUVs. Automatica **49**, 3359–3367 (2013)
8. Pang, Z., Zheng, C., Sun, J., Han, Q., Liu, G.: Distance- and velocity-based collision avoidance for time-varying formation control of second-order multi-agent systems. IEEE Trans. Circuits Syst. II, Exp. Briefs **68**(4), 1253–1257 (2021)
9. Sharma, R.S., Mondal, A., Behera, L.: Tracking control of mobile robots in formation in the presence of disturbances. IEEE Trans. Ind. Inf. **17**(1), 110–123 (2021)
10. Wang, B., Nersesov, S.G., Ashrafiuon, H.: Robust formation control and obstacle avoidance for heterogeneous underactuated surface vessel networks. IEEE Trans. Control Netw. Syst. **9**(1), 125–137 (2022)
11. Xia, G., Xia, X., Zheng, Z.: Formation tracking control for underactuated surface vehicles with actuator magnitude and rate saturations. Ocean Eng. **260**, 111935 (2022)

Chapter 6
Unit Modality Fast Docking Control in Large-Scale Environments

Abstract The segmented safe docking control framework for the fast assembly and docking of large-scale reconfigurable swarm in unit modalities, based on long-range global and short-range local perception, is proposed in this chapter. Through hierarchical safety constraints, adaptive anti-saturation, and the multiple extended state observer fusion disturbance estimation, the docking cascaded control framework achieves smoothness, robustness, and high-precision docking during the total combination reconfiguration process. The effectiveness of the proposed method is demonstrated through a series of simulations and experiments.

Keywords Combination reconfiguration · Segmented docking · Low-peak disturbance rejection

6.1 Introduction

Docking combination is the transition process from reconfigurable swarm unit to combined structure, which is key to its capability enhancement. Current unit modality combination docking mainly relies on high-precision sensor positioning under a single positioning reference (inertial system or vehicle system), sensing the unit modality and its surrounding environment, and driving multiple units to complete the combination. However, existing high-precision local sensing and docking methods for airborne systems [1–4] are not efficient in large-scale operation scenarios, and it is difficult to ensure that all swarm units are within each other's sensing range. The precision and quality of outdoor satellite navigation and indoor map global positioning used in actual applications cannot meet the precision constraints of the docking mechanism [5]. In addition, existing docking control of reconfigurable swarms mostly focuses on high-level strategies and motion planning. The lower-level control often adopts open-loop commands and simple kinematic closed-loop control, without considering application limitations such as performance saturation, environmental safety, docking process and sensing limitations. It also does not involve the

T. Jiang et al., *Multi-modality Control of Reconfigurable Swarm*,
https://doi.org/10.1007/978-981-95-0305-6_6

interference of cascaded kinematic errors and aggregate uncertainty on the upper-level kinematic tracking effect, thus its adaptability in actual complex environments is relatively poor.

Therefore, to address the shortcomings of existing reconfigurable swarm docking strategies and control methods in practical application scenarios, this book proposes a two-stage rapid combination docking control framework based on kinematic-dynamic joint structure for long-distance assembly and short-distance docking. In the long-distance stage, global positioning in the inertial coordinate system is used to provide the pose information of the unit modality and the docking target, and a kinematic controller is designed with saturation and safety constraints to drive the active docking unit modality to move to the position behind the docking target. This long-distance assembly does not require prior search and return, and can generate the desired assembly trajectory based on the global positioning of the unit modality and the docking target, quickly reaching the short-distance docking position. In the short-distance stage, considering the high precision requirements for docking mechanism locking, line-of-sight constraints, and limited docking travel, high-precision local sensing is relied upon to provide positioning information, and a short-distance kinematic controller is designed to drive the unit modality to quickly converge to the straight line guidance behind the docking target within the line-of-sight range, ensuring that the unit modality can deliver the lock precisely into the positioning hole of the docking target, completing reliable combination. At the dynamic layer, a new disturbance observation strategy with peak suppression capability is designed, namely the Multiple Extended State Observer (MESO) adaptive fusion estimation method to obtain low peak disturbance estimation signals, and feed them back to the bottom dynamic controller, achieving robust velocity tracking under unknown uncertainties. Finally, through the simulation and experiment of dynamic disturbance resistance and staged control, the adaptive advantages of the proposed method are demonstrated.

6.2 Problem Description

This chapter considers the combination of the ith unit and the kth unit for reconfiguration. Among them, the ith unit is the active docking unit, with a pose of $p_{ci} = [x_i, y_i, \theta_i]^{\mathrm{T}}$ in the inertial coordinate system. The kth unit is the passive docking unit, with a pose of $p_{ck} = [x_k, y_k, \theta_k]^{\mathrm{T}}$ in the inertial coordinate system.

The pose of the reconfigurable swarm unit in the inertial system is often obtained through satellite navigation or indoor mapping and positioning. The positioning accuracy is difficult to meet the precision requirements of the docking mechanism control. The end-to-end docking of the nearby reconfiguration combination is more suitable for high-precision local sensing. However, the range of local sensing is limited and it is difficult to achieve large-scale docking of multiple units. Therefore, a staged reconfiguration docking strategy is adopted. In the long distance, the inertial poses of the active and passive docking units generated continuously are used to design the tracking controller of the active docking unit, which moves to the position behind the positioning hole of the passive unit and ensures that the passive docking unit is within the local sensing range of the sensor of the active docking unit. In the short distance, based on the distance and direction sensed by the active unit to the passive docking target, it is converted into a high-precision pose under the stationary docking target coordinate system, and a tracking controller within the sensed line of sight is designed to deliver the lock of the active docking unit into the positioning hole, completing the locking and reconfiguration of the active and passive docking units. In the close-range phase, the pose of the active docking unit in the body inertial system is $p_{ci} = [x_i, y_i, \theta_i]^{\mathrm{T}}$, and in the passive docking unit coordinate system is $p_{ci}^k = \left[x_i^k, y_i^k, \theta_i^k\right]^{\mathrm{T}}$.

According to the unit modality motion model in Eq. (3.4), the corresponding kinematic control models are expressed as

$$
\begin{aligned}
\dot{x}_{il} &= v_i \cos(\theta_i) - l_i w_i \sin(\theta_i), \\
\dot{y}_{il} &= v_i \sin(\theta_i) + l_i w_i \cos(\theta_i), \\
\dot{\theta}_{il} &= w_i,
\end{aligned} \tag{6.1}
$$

$$
\begin{aligned}
\dot{x}_{il}^k &= v_i^k \cos(\theta_i) - l_i^k w_i^k \sin(\theta_i^k), \\
\dot{y}_{il}^k &= v_i^k \sin(\theta_i) + l_i^k w_i^k \cos(\theta_i^k), \\
\dot{\theta}_{il}^k &= w_i^k,
\end{aligned} \tag{6.2}
$$

where l_i is the axial offset between the center of gravity of the active docking unit and the center of the close-range phase measurement point, $x_{il} = x_i + l_i \cos(\theta_{il})$, $y_{il} = y_i + l_i \sin(\theta_{il})$ and $\theta_{il} = \theta_i$ are the poses of the active docking unit in the far-range phase considering the offset; l_i^k is the axial offset between the center of gravity of the active docking unit and the center of the far-range phase measurement point, similarly, $x_{il}^k = x_i^k + l_i^k \cos\left(\theta_{il}^k\right)$, $y_{il}^k = y_i^k + l_i^k \sin\left(\theta_{il}^k\right)$ and $\theta_{il}^k = \theta_i^k$ are the poses of the active docking unit in the close-range phase considering the offset.

This chapter still adopts the control framework of kinematics and dynamics coupling. Considering factors such as safety and energy efficiency, the amplitude and rate of linear velocity and angular velocity commands are subject to saturation constraints. The nested motion saturation models of velocity commands in the two phases are

$$
\begin{aligned}
\dot{v}_{is} &= \operatorname{sat}\left(-\bar{a}, \left(\operatorname{sat}\left(-\bar{v}, v_{id}, \bar{v}\right) - v_s\right)/\tau_v, \bar{a}\right), \\
\dot{w}_{is} &= \operatorname{sat}\left(-\bar{\omega}, \left(\operatorname{sat}\left(-\bar{w}, w_{id}, \bar{w}\right) - w_s\right)/\tau_w, \bar{\omega}\right),
\end{aligned} \tag{6.3}
$$

$$
\begin{aligned}
\dot{v}_{is}^k &= \operatorname{sat}\left(-\bar{a}, \left(\operatorname{sat}\left(-\bar{v}, v_{id}^k, \bar{v}\right) - v_s\right)/\tau_v, \bar{a}\right), \\
\dot{w}_{is}^k &= \operatorname{sat}\left(-\bar{\omega}, \left(\operatorname{sat}\left(-\bar{w}, w_{id}^k, \bar{w}\right) - w_s\right)/\tau_w, \bar{\omega}\right),
\end{aligned} \tag{6.4}
$$

In this, v_{id} and w_{id} are the linear velocity and angular velocity commands of the active docking unit in the close-range stage, and v_{is} and w_{is} are the corresponding commands for linear and angular velocities under nested saturation; v_{id}^k and w_{id}^k are the linear velocity and angular velocity commands of the active docking unit in the long-range stage, and v_{is} and w_{is} are the corresponding commands for linear and angular velocities under nested saturation.

In addition, in the long-range stage, the active docking unit is likely to encounter obstacle threats. If the position of the obstacle is (x_o, y_o), the position trajectory (x_i, y_i) in the long-range stage must ensure no interference with (x_o, y_o). In the close-range stage, it is necessary to ensure that the docking trajectory $\left(x_i^k, y_i^k\right)$ quickly converges to the desired trajectory, and the posture is always within the line of sight, that is, $|\theta_i^k| < \theta_{sf}$, where $\theta_{sf} \in \mathbf{R}_+$ is determined by the sensing range of the onboard sensor.

Based on the above analysis, the linear velocity and angular velocity commands after nested saturation constraints can be converted into wheel velocity commands ω_{ild} and ω_{ird} for the active docking unit. The wheel velocity command is the reference input for the dynamic control model. As known from the nested saturation constraint, the wheel velocity command and its first derivative are bounded. The dynamic control model used by the active docking unit is given by Eq. (4.2).

The overall analysis of the safety reconfiguration control of the unit modality at the kinematic level is shown in Fig. 6.1. Using the global positioning of the inertial system and the carrier sensing of the docking target system, it is divided into long-range assembly and close-range docking stages. Combining multi-dimensional constraints, given control elements and control objectives in different stages, a stage-by-stage combined docking controller is designed, as follows:

(1) In the far-field stage, the rendezvous trajectories x_d and y_d and their first and second derivatives are bounded. Considering the control model (6.1) under the nested saturation constraint (6.3), we design the rendezvous trajectory control commands u_{ix} and u_{iy}, so that the active docking unit can bypass the obstacle (x_o, y_o) and move smoothly and safely to the rendezvous point.

(2) In the near-field stage, the docking trajectories x_d^k and y_d^k and their first and second derivatives are bounded. Considering the control model (6.2) under the nested saturation constraint (6.4), we design the docking trajectory control commands u_{ix}^k and u_{iy}^k, so that the active docking unit can converge steadily and quickly to the guidance trajectory within the line-of-sight safety range $[\theta_{sr}, \theta_{sl}]$, completing the reconfiguration of the lock and positioning hole.

This book adopts the kinematics and dynamics level control, and the purpose of the lower level dynamics control is to design a velocity tracking controller based on the model (4.2), to achieve robust velocity tracking control and disturbance peak suppression of the active docking unit under unknown aggregate uncertainty.

The following are the related lemmas of the control safety and stability in this book.

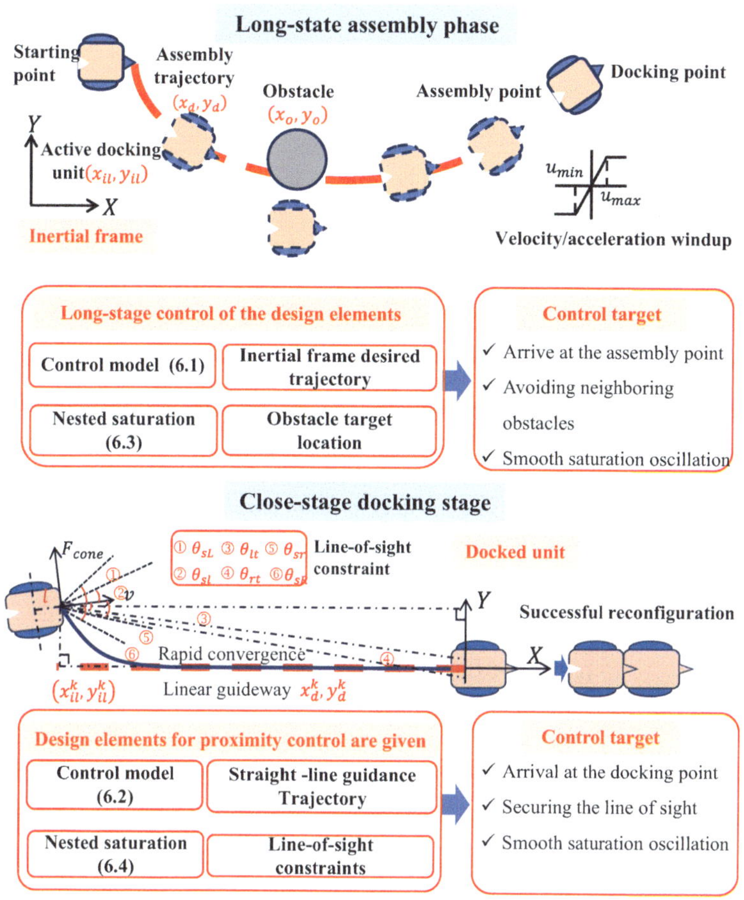

Fig. 6.1 Sectional safety docking control block diagram

Lemma 6.1 *If there exist the following double integral system, obstacle function, safe acceleration envelope, and control law*

$$\dot{\eta}_1 = \eta_2, \quad \dot{\eta}_2 = u, \tag{6.5}$$

$$\mathcal{B}(\eta_1, \eta_2) = \eta_2 - \mathcal{F}(\eta_1), \tag{6.6}$$

$$u_s(\eta_1, \eta_2) = -\alpha\mathcal{B}(\eta_1, \eta_2) + \mathcal{H}(\eta_1)\eta_2, \tag{6.7}$$

$$u = sat(-\bar{u}, \min\{u_r(\eta, \eta_r), u_s(\eta_1, \eta_2)\}, \bar{u}), \tag{6.8}$$

where,

$$h(x) = \begin{cases} 1, & |x| \geq 0, \\ 0, & |x| < 0, \end{cases}$$

$$\beta(x) = \left[2\gamma\rho \ln\left(\cosh\left(\frac{x}{\rho}\right)\right)\right]^{\frac{1}{2}}, \quad x \leq 0,$$

$$\mathcal{H}(x) = h(-x) \frac{\gamma \tanh\left(\frac{x}{\rho}\right)}{\beta(x)} \sqrt{\frac{\gamma}{\rho}},$$

$$\mathcal{F}(x) = -h(-x)\beta(x) + (1 - h(-x)) \sqrt{\frac{\gamma}{\rho}} x,$$

then the designed gains γ, ρ and α satisfy $0 < \gamma < \bar{u}$, $\rho > 0$ and $\alpha > 0$, where $\bar{u} > 0$ is the saturation threshold. The to-be-designed control law $u_r(\eta, \eta_r)$ generates the actual control law (6.8) under the safe acceleration envelope (6.7), which can ensure that the system position state η_1 is always within the set safe boundary [6].

Lemma 6.2 *Consider the following perturbed time-varying system*

$$\dot{\xi}(t) = \frac{1}{\epsilon} \mathcal{K}(t) \xi(t) + \delta(t), t \geq t_0, t_0 \in \boldsymbol{R}_+, \tag{6.9}$$

where $\mathcal{K} \in C(\boldsymbol{R}_{\geq t_0})$, $\delta(t)$ is the disturbance term, and the initial condition $\xi(t_0)$ belongs to a compact subset of $\boldsymbol{R}$. Suppose the linear time-varying system $\dot{\xi}_0(t) = \mathcal{M}(t)\xi_0(t)$ is uniformly exponentially stable, and $|\delta(t)| \leq \bar{\delta}/(\varepsilon^m)$, where $\bar{\delta}$ is a positive constant independent of ε, and $m \geq 0$ is an integer. Then for any $\sigma > 0$ and $T > 0$, there exists $\varepsilon_0 > 0$ such that for any $\varepsilon \in (0, \varepsilon_0)$ and $t \in [t_0 + T, \infty)$, the solution of Eq. (6.9) satisfies $\|\xi(t)\| \leq \sigma\varepsilon^m$ [7].

6.3 Adaptiver Fault-Tolerant Handover Strategy Design

Reconfigurable swarm units can complete the transition from individual units to a combined entity through reconfiguration. When a unit in coordination encounters complex terrain or other sudden tasks that require reconfiguration into a new combination configuration, it is difficult to guarantee the existence of suitable docking units within the local sensing range; especially when the two units to be combined are far apart, the efficiency of the preliminary search behavior of the existing docking strategy cannot effectively respond to task demands. Therefore, combining sensing characteristics and task characteristics, the global positioning coordinates of each unit in the network can be used to publish suitable assembly point positions, navigate the active docking units to assemble to the rear of the target to be docked within the local sensing range, and then complete the locking combination by delivering the

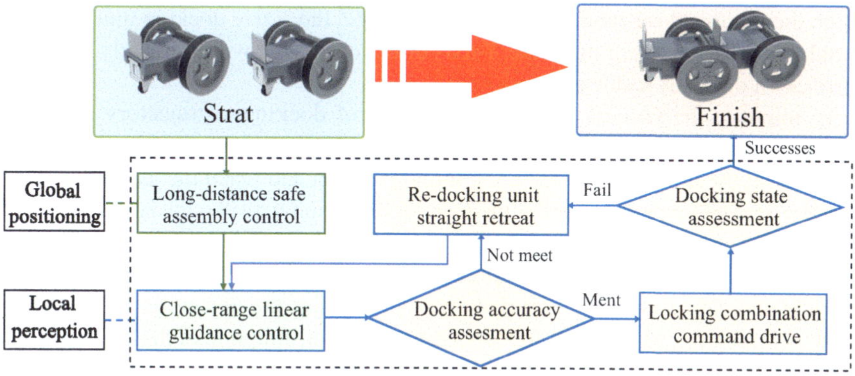

Fig. 6.2 Segmented fault-tolerant docking process

lock into the positioning hole through straight-line guidance. At the same time, to improve the success rate of docking and fault tolerance, a re-docking mechanism can be introduced based on the judgment of docking accuracy and locking combination status, ensuring complete reconfiguration between units.

The specific task flow diagram is shown in Fig. 6.2. In complex scenarios, the unit modality may encounter obstacles during the assembly process in the long-distance docking stage. Considering the real-time existence of control input saturation of the unit modality, the controller design must involve adaptive anti-saturation smooth reaction obstacle avoidance to ensure the safety of assembly. In the close-range stage, due to the errors in global positioning using satellite positioning or indoor mapping and other inertial positioning methods, the unit modality after assembly will inevitably deviate from the rear of the docking target, which can easily trigger subsequent control input saturation and trajectory oscillation. Therefore, the design of the close-range controller needs to combine adaptive anti-saturation and line-of-sight safety maintenance to achieve smooth straight-line guidance movement within the sensing range. When docking fails, the unit modality that actively docks will retreat along the straight line of the docking target axis and re-enter the close-range straight-line guidance control stage to ensure complete reconfiguration.

6.4 Long-Distance Assembly Safe Trajectory Planning

In complex operational scenarios, units are prone to encounter various types of obstacles during the process of long-distance assembly. Incorporating a smooth reactive obstacle avoidance mechanism into the controller design can ensure that the unit safely reaches the assembly point. However, the posture of the unit when it reaches the assembly point is not constrained. This means that if the active docking unit has a large angular deviation from the target to be docked when it arrives at the assembly point, then in the close-range linear guidance stage, the active docking unit needs to spend more time converging in the Y-axis direction, which will ultimately lead to a decrease in docking accuracy.

To improve the success rate and robustness of docking, a trajectory planning method can be used to design the docking trajectory. By applying constraints, it ensures that the unit can start from the initial position and reach the assembly point with the desired posture, reducing the angular deviation between the two units to be combined. Thus, in the close-range linear guidance stage, the active docking unit can quickly converge in the Y-axis direction. Reactive obstacle avoidance, as a local obstacle avoidance method, carries the risk of causing the unit to oscillate during trajectory tracking, which leads to increased energy consumption of the unit, reduced trajectory tracking performance, and increased avoidance costs. To reduce the cost of the unit avoiding obstacles, the existence of static obstacles can be considered in the docking trajectory planning, and a collision-free safe docking trajectory can be planned. In addition, to further ensure the safety of the unit assembly, the reactive obstacle avoidance mechanism is still applied in the controller design.

6.4.1 Multi-constraint Trajectory Planning Based on Differential Evolution Algorithm

In order to satisfy the motion characteristics of the unit modality, the planned trajectory and corresponding velocity and acceleration (or higher order derivatives) need to be continuous. Polynomials naturally satisfy these continuity and differentiability requirements, so they can be used to generate trajectories for the X and Y axes, and then calculate the corresponding linear velocity and angular velocity through the kinematic equations of the moving unit. To ensure that the polynomial trajectory meets environmental and kinematic constraints and guarantees the safe assembly of the unit modality, the differential evolution algorithm is used to optimize the polynomial coefficients of the trajectory to obtain the optimal long-distance safe assembly trajectory.

To simplify the optimization variables and set constraints, a fourth-order polynomial about time is used here to describe the trajectories of the X and Y axes. For simplicity, the trajectories of the X and Y axes are uniformly described as $f(t)$, which is expressed as follows

$$f(t) = p_0 + p_1 t + p_2 t^2 + p_3 t^3 + k t^4 \tag{6.10}$$

where p_0, p_1, p_2, p_3, k are the parameters of the trajectory.

Taking the derivative with respect to time, we can get the expression for velocity as

$$\dot{f}(t) = p_1 + 2p_2 t + 3p_3 t^2 + 4kt^3 \tag{6.11}$$

Set the initial position $f(0)$, initial velocity $\dot{f}(0)$, final position $f(T)$, final velocity $\dot{f}(T)$, and after obtaining the final time T, according to the expressions of trajectory and velocity, the following equations can be established

$$\begin{bmatrix} 1 & 0 & 0 & 0 \\ 0 & 1 & 0 & 0 \\ 1 & T & T^2 & T^3 \\ 0 & 1 & 2T & 3T^2 \end{bmatrix} \begin{bmatrix} p_0 \\ p_1 \\ p_2 \\ p_3 \end{bmatrix} = \begin{bmatrix} f(0) \\ \dot{f}(0) \\ f(T) - kT^4 \\ \dot{f}(T) - 4kT^3 \end{bmatrix} \tag{6.12}$$

If the trajectory parameter k is determined and the final time satisfies $T > 0$, then the other parameters of the trajectory p_0, p_1, p_2, p_3 will also be uniquely determined.

So far, the trajectory can be completely described. The specific expressions of the X-axis, Y-axis trajectories and velocities are as follows

$$\begin{cases} x = f_x(t) = p_{0x} + p_{1x}t + p_{2x}t^2 + p_{3x}t^3 + k_x t^4 \\ y = f_y(t) = p_{0y} + p_{1y}t + p_{2y}t^2 + p_{3y}t^3 + k_y t^4 \end{cases} \tag{6.13}$$

$$\begin{cases} \dot{x} = p_{1x} + 2p_{2x}t + 3p_{3x}t^2 + 4k_x t^3 \\ \dot{y} = p_{1y} + 2p_{2y}t + 3p_{3y}t^2 + 4k_y t^3 \end{cases} \tag{6.14}$$

Here, $p_{ix}(i = 0 \sim 3)$ and k_x are the parameters of the X-axis trajectory; $p_{iy}(i = 0 \sim 3)$ and k_y are the parameters of the Y-axis trajectory.

According to the kinematic model of the differential drive vehicle, the linear velocity, angular velocity, and pose information can be extracted from the trajectory and velocity in the following way:

$$\begin{cases} v = \sqrt{(\dot{x})^2 + (\dot{y})^2} \\ \theta = \arctan(\dfrac{\dot{y}}{\dot{x}}) \\ \omega = \dfrac{\ddot{y}\dot{x} - \ddot{x}\dot{y}}{\dot{x}^2 + \dot{y}^2} \end{cases} \tag{6.15}$$

When $t = T$, the calculated pose $\omega(T)$ is the desired pose ω_d when the active docking unit reaches the remote assembly point. It is worth noting that to prevent singularities when calculating the desired pose at the remote assembly point, the linear velocity at the end of the trajectory should be greater than zero and less than the threshold that can drive the unit modality movement. This will not cause displacement of the unit modality, nor will it cause singularities in angle calculation. In addition, to ensure the success rate of docking, the desired pose should be consistent with the direction of the target to be docked, to accelerate the convergence velocity of the Y-axis in the subsequent straight-line guidance stage. Finally, combining the linear velocity threshold v_{th} of the unit modality movement and the direction θ_f of the target to be docked, the components of the velocity along the X-axis $\dot{x}_f$ and the

Y-axis $\dot{y}_f$ at the end of the trajectory can be derived, as shown below:

$$\begin{cases} \dot{x}_f = v_{th} \cos \theta_f \\ \dot{y}_f = v_{th} \sin \theta_f \end{cases} \tag{6.16}$$

To ensure that the active docking unit can reach the assembly point in the shortest time with the desired posture, many constraints need to be applied to the trajectory planning. For this kind of multi-constraint, multi-variable optimization problem, conventional optimization algorithms are prone to fall into local optima and are difficult to handle, so the differential evolution algorithm is used for solving. To make the trajectory satisfy all kinds of constraints, it is necessary to add the constraint conditions to the calculation of the objective function. To calculate the objective function of each individual, it is necessary to discretize the trajectory. Here, the trajectory points are discretized into N points, and the sampling time intervals are the same.

Considering the kinematic constraints, it is specifically manifested as the existence of upper limits for the linear velocity and angular velocity of the unit modality. To prevent differential evolution from falling into local optima during the solving process, the objective function must continuously change during the search process. The expression of the objective function for the part of the kinematic constraint is as follows

$$f_\omega = \begin{cases} 0 & abs(\omega) \le \omega_s \\ 1 - \frac{1}{1+k_\omega(abs(\omega)-\omega_s)} & abs(\omega) > \omega_s \end{cases} \tag{6.17}$$

$$f_v = \begin{cases} 0 & abs(v) \le v_s \\ 1 - \frac{1}{1+k_v(abs(v)-v_s)} & abs(v) > v_s \end{cases} \tag{6.18}$$

where, v_s and ω_s represent the maximum upper limits of linear velocity and angular velocity respectively; k_v and k_ω are gains, determining the rate of change of the objective functions f_v and f_ω. Consider the environmental constraints. To avoid the active docking unit colliding with obstacles during the assembly process, an Artificial Potential Field (APF) is introduced in the construction of the objective function. The repulsive potential field part of it is continuous with respect to the relative distance between the unit modality and the obstacle, which can effectively avoid the algorithm falling into local optimum. At the same time, when the relative distance tends to 0, the repulsive force tends to infinity, ensuring the safety of the active docking unit. The magnitude of the repulsive force experienced at any point (x_a, y_a) on the trajectory is shown as follows

$$d_{rel} = \sqrt{(x_a - x_{obs})^2 + (y_a - y_{obs})^2} \tag{6.19}$$

$$f_{rep} = \begin{cases} 0 & d_{rel} > d_s \\ (\frac{1}{d_{rel}} - \frac{1}{d_s})(\frac{1}{d_s})^2 & d_{rel} \le d_s \end{cases} \tag{6.20}$$

where (x_{obs}, y_{obs}) is the position coordinates of the obstacle; d_s is the radius of the sensed obstacle.

Based on this, the objective function for the collision avoidance part can be described as

$$f_{ca} = \sum_{i=0}^{N} f_{rep,i} \Delta t \tag{6.21}$$

In summary, the final total objective function expression is as follows

$$f_{obj} = w_1 f_\omega + w_2 f_v + w_3 f_{ca} + w_4 T \tag{6.22}$$

where $w_1 \sim w_4$ are the weights of each sub-objective function.

From the above analysis, it can be seen that the size of the objective function depends on the coefficients of the trajectory polynomial and the end time T of the trajectory. The polynomial coefficients are uniquely determined by the trajectory parameters k_x, k_y and the end time T. Therefore, k_x, k_y and T can be considered as optimization variables, and each individual in the differential evolution algorithm must include these elements. The differential evolution algorithm iterates to minimize the objective function, obtains the optimal individual, and thus solves for an optimal long-distance docking safe trajectory.

6.5 Wheel Velocity Dynamic Robust Control Based on Low-Peak Disturbance Estimation

6.5.1 *Adaptive Fusion Disturbance Estimation Based on Multiple Extended State Observers*

(1) Design of Multi-ESO Adaptive Fusion Disturbance Estimation Scheme

Consider the following general uncertain nonlinear system model

$$\begin{cases} \dot{x}_{(1)}(t) = x_{(2)}(t), \\ \dot{x}_{(2)}(t) = x_{(3)}(t), \\ \quad \vdots \\ \dot{x}_{(n)}(t) = bu(t) + d(t), \\ y(t) \quad = x_{(1)}(t), \end{cases} \tag{6.23}$$

where $x(t) \triangleq [x_{(1)}(t), \ldots, x_{(n)}(t)]^T \in \mathcal{X} \subseteq \mathbf{R}^n$, $d(t) \in \mathcal{D} \subseteq \mathbf{R}$ and $u(t) \in \mathcal{U} \subseteq \mathbf{R}$ are the system state, aggregated uncertainty, and control input, respectively.

For system (6.23), the extended state $x_{n+1}(t) \triangleq d(t) \in \mathcal{D} \subseteq \mathbf{R}$ can be considered as the aggregated uncertainty; $h(t) \triangleq \dot{x}_{n+1}(t)$ is the derivative of the extended state. The matrix form of the uncertain nonlinear system is written as follows

$$\dot{x}(t) = Ax(t) + Bbu(t) + Dh(t),\ y(t) = Cx(t), \tag{6.24}$$

and the matrices are

$$A \triangleq \begin{bmatrix} 0_{n\times 1} & I_{n\times n} \\ 0_{1\times 1} & 0_{1\times n} \end{bmatrix},\ B \triangleq \begin{bmatrix} 0_{1\times(n-1)} & I_{1\times 1} & 0_{1\times 1} \end{bmatrix}^{\mathrm{T}},$$
$$C \triangleq \begin{bmatrix} I_{1\times 1} & 0_{1\times n} \end{bmatrix},\ D \triangleq \begin{bmatrix} 0_{1\times(n-1)} & 0_{1\times 1} & I_{1\times 1} \end{bmatrix}^{\mathrm{T}},$$

where $x(t) \triangleq \left[x_{(1)}(t), \ldots, x_{(n+1)}(t)\right]^{\mathrm{T}} \in \mathcal{X}' \subseteq \mathbf{R}^{n+1}$ are the extended states, including the system state and unknown disturbance.

The block diagram of the proposed method is shown in Fig. 6.3. The system state and aggregated uncertainty are reconstructed through the fusion estimation of N ESOs, where $N = n + 2$, similar to the quantity definition method proposed by Shakarami et al. [8]. The design of a single ESO is as follows

$$\begin{aligned} \dot{\hat{x}}_i(t) &= A\hat{x}_i(t) + Bbu(t) + H(y(t) - \hat{y}_i(t)), \\ \hat{y}_i(t) &= C\hat{x}_i(t), \end{aligned} \tag{6.25}$$

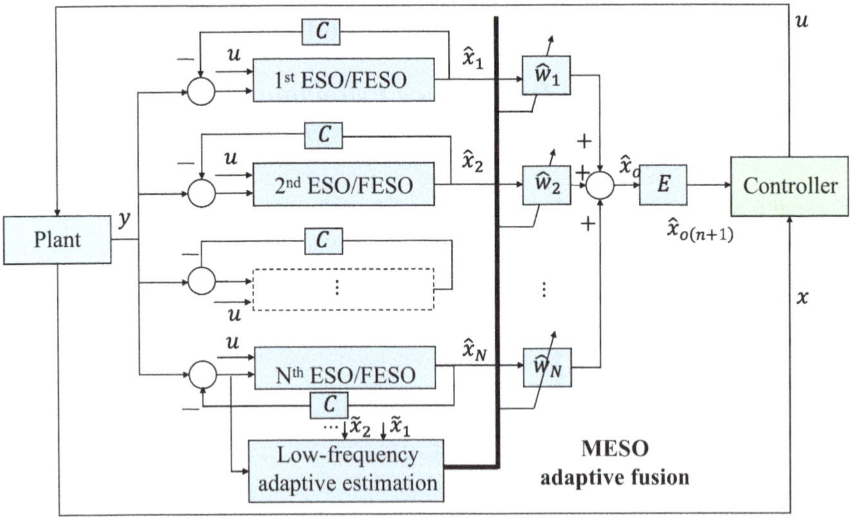

Fig. 6.3 Adaptive fusion estimation framework of MESO

where $\hat{x}_i \triangleq \left[\hat{x}_{i(1)}(t), \ldots, \hat{x}_{i(n)}(t), \hat{x}_{i(n+1)}(t)\right]^{\mathrm{T}} \in \mathbf{R}^{n+1}$ is the estimate of the state and aggregated uncertainty from the ith ESO, where $\hat{x}_{i(j)}(t)$ represents the jth element of the ith ESO, $H \triangleq \left[\frac{\kappa_1}{\epsilon}, \ldots, \frac{\kappa_{n+1}}{\epsilon^{n+1}}\right] \in \mathbf{R}^{n+1}$.

The fusion of multiple ESOs can be written as follows:

$$\hat{x}_o(t) = \sum_{i=1}^{N} w_i \hat{x}_i(t), \sum_{i=1}^{N} w_i = 1, \tag{6.26}$$

where the fusion weight of the ith ESO is $w_i \in [0, 1]$, $\hat{x}_o(t) \in \left[\hat{x}_{o(1)}(t), \ldots, \hat{x}_{o(n)}(t), \hat{x}_{o(n+1)}(t)\right]^{\mathrm{T}} \in \mathbf{R}^{n+1}$, $\hat{x}_{o(j)}(t)$ represents the jth element of $\hat{x}_o(t)$, which is the final fusion result of MESO.

Subtracting Eq. (6.25) from Eq. (6.24), we get

$$\dot{\tilde{x}}_i(t) = A_0 \tilde{x}_i(t) + Eh(t) - HDv(t), \tag{6.27}$$

where $A_0 \triangleq A - HC$, $\tilde{x}_i(t) \triangleq x(t) - \hat{x}_i(t)$ represents the estimation error of the ith ESO.

There exists an unknown weight $w_i^{\star}$ that achieves the optimal fusion of multiple ESOs

$$x(t) = \sum_{i=1}^{N} w_i^{\star} \hat{x}_i(t) + \varepsilon(t), \forall t > 0, \tag{6.28}$$

where $\varepsilon(t) \in \mathbf{R}^{n+1}$ is the approximation error caused by uncertain dynamics, and $\|\varepsilon(t)\| \leq \varepsilon_d$ and $\|\varepsilon(0)\| = 0$. Note that ε can be equivalent to the estimation error of the steady-state phase of a single ESO.

Rearranging and rewriting Eq. (6.28), we get

$$\sum_{i=1}^{N-1} w_i^{\star}(\tilde{x}_i(t) - \tilde{x}_N(t)) - \varepsilon(t) = -\tilde{x}_N(t). \tag{6.29}$$

The recursive form of Eq. (6.29) can be written as

$$\Psi(t) W^{\star} - \varepsilon(t) = -\tilde{x}_N(t), \tag{6.30}$$

where $\Psi(t) \triangleq \left[\tilde{x}_1(t) - \tilde{x}_N(t), \ldots, \tilde{x}_{N-1}(t) - \tilde{x}_N(t)\right]$ is the completely known error of multiple ESOs, $W^{\star} \triangleq \left[w_1^{\star}, \ldots, w_{N-1}^{\star}\right]^{\mathrm{T}}$ is the unknown constant weight vector.

In order to obtain the estimate of $W^{\star}$, we use low-frequency adaptive learning technology to estimate the weight vector, as shown below

$$
\begin{aligned}
\dot{\hat{W}}(t) &= -K(t)\Psi^{\mathrm{T}}(t)C^{\mathrm{T}}(C\Psi(t)\hat{W}(t) + \tilde{y}_N(t)) \\
&\quad - \sigma K(t)\Psi^{\mathrm{T}}(t)C^{\mathrm{T}}(C\Psi(t)(\hat{W}(t) - \hat{W}_f(t))), \\
\dot{K}(t) &= -K(t)\Psi^{\mathrm{T}}(t)C^{\mathrm{T}}C\Psi(t)K(t), \\
\dot{\hat{W}}_f(t) &= \tau_f(\hat{W}(t) - \hat{W}_f(t)),
\end{aligned}
\tag{6.31}
$$

In this, $\tilde{y}_N(t) \triangleq y(t) - \hat{y}_N(t)$; $\hat{y}_N(t) \triangleq C\hat{x}_N(t)$; $\hat{W} \in \mathbf{R}^{N-1}$ is an estimate of $W^{\star}$; $\hat{W}_f \in \mathbf{R}^{N-1}$ is the low-pass filter weight estimate of $\hat{W}$; $K(t) \in \mathbf{R}^{(N-1)\times(N-1)}$ is the variable gain matrix, where $K(0) = \gamma I^{(N-1)\times(N-1)}$, γ is a positive number, and $\tau_f \in \mathbf{R}^{+}$ is the positive filter gain. For convenience of expression, all time-related variables in this book, such as $x(t)$ and $y(t)$, will hide the time t when they appear later.

Through the estimation of the weight parameter $\hat{W}$ in (6.31), MESO adaptive fusion design is used to estimate the aggregate uncertainty, including unknown nonlinear dynamics and external disturbances. It is worth noting that $\hat{x}_{o(n+1)}$ is the estimated value of the aggregate uncertainty.

Assumption 6.1 The initial conditions of ESO (6.25) are chosen so that the matrix $\Psi(0)\Psi(0)^{\mathrm{T}}$ is full rank, and each ESO's A_0 is a Hurwitz matrix.

Remark 6.1 Similar to the low-frequency learning law proposed by Yucelen et al. [9], γ determines the adaptive learning rate. A higher γ value can make the adaptive fusion weight converge quickly, but it may also cause inherent high-frequency oscillations, leading to system instability. The third differential equation in (6.31) generates the filtered and smoothed adaptive fusion weights. By coordinating the updated adaptive fusion weight deviation and the adjustment factor σ, a compensation term in the first differential equation in (6.31) is formed, which can be used for the adaptive update of the fusion weight. This measure helps to improve the smoothness and stability margin of the MESO adaptive fusion estimate.

Remark 6.2 Assume 4.1 can be found in the bounded precondition proposed by Ran et al. [7]. In practical applications, these states and variables are not only reasonable but also easy to satisfy. Assume the condition $\Psi(0)\Psi(0)^{\mathrm{T}}$ in Assumption 4.2 is full rank, which can be achieved by setting the initial value of each ESO. Furthermore, $\Psi(0)\Psi(0)^{\mathrm{T}}$ is full rank and has an inverse matrix, which is a necessary precondition for subsequent response analysis and stability proof. Assume A_0 is a Hurwitz matrix, indicating that a single ESO is stable. This is an easy-to-achieve precondition, and the required steady-state and transient estimation performance can be obtained by adjusting the ESO gain [10].

(2) Analysis of disturbance estimation steady-state and transient performance

Consider the proportional estimation error of MESO fusion and a single ESO $\eta \triangleq \left[\eta_{(1)}, \ldots, \eta_{(n+1)}\right]^{\mathrm{T}} \in \mathbf{R}^{n+1}$, where $\eta_{(i)} \triangleq \left(x_{(i)} - \hat{x}_{o(i)}\right)/\epsilon^{n+1-i} \in \mathbf{R}$ and $\eta_i \triangleq G(\epsilon)^{-1}\left(x - \hat{x}_i\right) \in \mathbf{R}^{n+1}$, $G(\epsilon) \triangleq \mathrm{diag}(\epsilon^{n}, \ldots, \epsilon, 1) \in \mathbf{R}^{(n+1)\times(n+1)}$. The error dynamics of MESO fusion and low-frequency adaptive learning law can be written as

$$\eta = \Psi_1 \hat{W} + \eta_N,$$
$$\dot{\hat{W}} = -\epsilon^{2n} K \Psi_1^{\mathrm{T}} C^{\mathrm{T}} C \eta - \epsilon^n K \Psi_1^{\mathrm{T}} C^{\mathrm{T}} D - \sigma \epsilon^{2n} K \Psi_1^{\mathrm{T}} C^{\mathrm{T}} C \Psi_1 \left(\hat{W} - \hat{W}_f\right), \quad (6.32)$$

where $\Psi_1 \triangleq \left[\eta_1 - \eta_N, \ldots, \eta_{N-1} - \eta_N\right] \in \mathbf{R}^{(n+1)\times(N-1)}$, $\sigma \in \mathbf{R}^+$ is a design parameter.

The derivatives of η, η_i and Ψ_1 can be written as

$$\dot{\eta}_i = \frac{1}{\epsilon} A_1 \eta_i + D\dot{d}, \quad (6.33)$$

$$\dot{\eta} = -\epsilon^{2n} \Psi_1 K \Psi_1^{\mathrm{T}} C^{\mathrm{T}} C \eta + E\dot{d} + \frac{1}{\epsilon} A_1 \eta - \sigma \Psi_1 \epsilon^{2n} K \Psi_1^{\mathrm{T}} C^T C \Psi_1 \left(\hat{W} - \hat{W}_f\right), \quad (6.34)$$

$$\dot{\Psi}_1 = \frac{1}{\epsilon} A_1 \Psi_1, \quad (6.35)$$

where $A_1 \triangleq \epsilon G(\epsilon)^{-1} A_0 G(\epsilon) \in \mathbf{R}^{(n+1)\times(n+1)}$ and $H_1 \triangleq \left[\kappa_1, \ldots, \kappa_{n+1}\right]^{\mathrm{T}} \in \mathbf{R}^{n+1}$. By integrating equation (6.35), we have $\Psi_1 = e^{\frac{1}{\epsilon} A_1 t} \Psi_1(0)$, which is a Hurwitz matrix.

Furthermore, we choose the candidate Lyapunov function $V_\Psi = tr\left(e^{\frac{1}{\epsilon} A_1^{\mathrm{T}} t} P e^{\frac{1}{\epsilon} A_1 t}\right)$, where the positive definite matrix P satisfies $A_1^{\mathrm{T}} P + P A_1 = -I$, and the corresponding derivative is written as $\dot{V}_\Psi = -\frac{1}{\epsilon} tr\left(e^{\frac{1}{\epsilon} A_1^{\mathrm{T}} t} e^{\frac{1}{\epsilon} A_1 t}\right)$. Then, we have

$$V_\Psi \leq e^{-\frac{1}{\epsilon \lambda_{\max}(P)} t} V_\Psi(0), \quad (6.36)$$

$$\left\| e^{\frac{1}{\epsilon} A_1 t} \right\|^2 \leq tr\left(e^{\frac{1}{\epsilon} A_1^{\mathrm{T}} t} e^{\frac{1}{\epsilon} A_1 t}\right) \leq (n+1) \left\| e^{\frac{1}{\epsilon} A_1 t} \right\|^2, \quad (6.37)$$

$$\left\| e^{\frac{1}{\epsilon} A_1 t} \right\|^2 \leq \frac{V_\Psi}{\lambda_{\min}(P)} \leq \frac{\lambda_{\max}(P)}{\lambda_{\min}(P)} \left\| e^{\frac{1}{\epsilon} A_1 t} \right\|^2, \quad (6.38)$$

where $\lambda_{\min}(P)$ and $\lambda_{\max}(P)$ are the minimum and maximum eigenvalues of P, respectively. Therefore, we can obtain the following inequality

$$\left\| e^{\frac{1}{\epsilon} A_1 t} \right\| \leq k_0 e^{-\frac{\lambda_0}{\epsilon} t}, \quad (6.39)$$

where $k_0 \triangleq \sqrt{\frac{(n+1)\lambda_{\max}(P)}{\lambda_{\min}(P)}} \in \mathbf{R}^+$, $\lambda_0 \triangleq \frac{1}{2\lambda_{\max}(P)} \in \mathbf{R}^+$.

Theorem 6.1 *Consider the uncertain nonlinear system (6.23) and the proposed adaptive MESO fusion (6.25), (6.26) and (6.31). Assume that Assumptions 4.1–4.2 are satisfied, and the initial conditions of the MESO fusion belong to a compact subset of* $\mathbf{R}^{n+1}$. *Then there exist* $\epsilon > 0$ *and* $\gamma > 0$ *such that* $\|\tilde{x}_o\|$, $\|\Psi_1 \tilde{W}\|$ *and* $\|\Psi_1 \tilde{W}_f\|$ *have* *bounds dependent on ϵ and γ, and the peak of the estimated signal can be adjusted and suppressed by ϵ, μ, σ, and τ_f.*

Proof Equation (6.30) can be used to obtain $\Psi_1 W^\star + \eta_N(W^\star) - G^{-1}(\epsilon)\varepsilon = 0$. By using Eq. (6.32), we have

$$\eta = \Psi_1 \tilde{W} + \varepsilon_1, \tag{6.40}$$

where $\tilde{W} \triangleq \hat{W} - W^\star \in \mathbf{R}^{(N-1)\times(N-1)}$, $\varepsilon_1 \triangleq G^{-1}(\epsilon)\varepsilon \in \mathbf{R}^{n+1}$.

The derivative of the estimation error of the MESO fusion weight is written as

$$\dot{\tilde{W}} = -\sigma\epsilon^{2n} K \Psi_1^{\mathrm{T}} C^{\mathrm{T}} C \left(\Psi_1 \left(\hat{W} - \hat{W}_f\right)\right) - \epsilon^{2n} K \Psi_1^{\mathrm{T}} C^{\mathrm{T}} C \left(\Psi_1 \tilde{W} + \varepsilon_1\right). \tag{6.41}$$

Combining $\dfrac{dK^{-1}\tilde{W}}{dt} = -K^{-1}\dot{K}K^{-1}\tilde{W} + K^{-1}\dot{\tilde{W}}$ and Eq. (6.41), the derivative of $K^{-1}\tilde{W}$ is as follows

$$\frac{dK^{-1}\tilde{W}}{dt} = -\sigma\epsilon^{2n}\Psi_1^{\mathrm{T}} C^{\mathrm{T}} C \left(\Psi_1 \left(\hat{W} - \hat{W}_f\right)\right) - \epsilon^{2n}\Psi_1^{\mathrm{T}} C^{\mathrm{T}} C\varepsilon_1. \tag{6.42}$$

Integrating equation (6.42), we can obtain

$$\begin{aligned} \tilde{W} = {} & K K^{-1}(0)\,\tilde{W}(0) - \sigma\epsilon^{2n} K \int_0^t \Psi_1^{\mathrm{T}} C^{\mathrm{T}} C \Psi_1 \left(\hat{W} - \hat{W}_f\right) d\tau \\ & - \epsilon^{2n} K \int_0^t \Psi_1^{\mathrm{T}} C^{\mathrm{T}} C \varepsilon_1 d\tau. \end{aligned} \tag{6.43}$$

Integrating $\dfrac{dK^{-1}}{dt} = -K^{-1}\dot{K}K^{-1}$, K can be expressed as

$$K = \left[K^{-1}(0) + \epsilon^{2n}\Psi_1^{\mathrm{T}}(0)\,\Phi_o(t)\Psi_1(0)\right]^{-1}, \tag{6.44}$$

$$K = K(0) - \Psi_1^{\mathrm{T}}(0) M_1 \Psi_1(0) K(0), \tag{6.45}$$

where

$$M_1 = \left(\frac{1}{\gamma\epsilon^{2(n)}}\Phi_o^{-1}(t) + \Psi_1(0)\Psi_1^{\mathrm{T}}(0)\right)^{-1},$$

and the symmetric and positive definite matrix

$$\Phi_o \triangleq \int_0^t e^{\frac{1}{\epsilon}A_1^{\mathrm{T}}\tau} C^{\mathrm{T}} C e^{\frac{1}{\epsilon}A_1\tau} d\tau \in \mathbf{R}^{(n+1)\times(n+1)}.$$

Then, according to the Neumann series, we have

$$\left(I+\frac{1}{\gamma\epsilon^{2(n)}}\Phi_o^{-1}(t)\left(\Psi_1(0)\Psi_1^{\mathrm{T}}(0)\right)^{-1}\right)^{-1}=\sum_{k=0}^{\infty}\frac{(-1)^k}{\mu^k}\left(\Phi_o^{-1}(t)\left(\Psi_1(0)\Psi_1^{\mathrm{T}}(0)\right)^{-1}\right)^k, \tag{6.46}$$

where $\mu \triangleq \gamma\epsilon^{2n} \in \mathbf{R}^+$ is a positive constant.

Through Eqs. (6.32), (6.43) and (6.46), we have

$$\Psi_1\tilde{W}=e^{\frac{1}{\epsilon}A_1t}G_1\Psi_1(0)\tilde{W}(0)-e^{\frac{1}{\epsilon}A_1t}\Phi_o^{-1}G_2M_2, \tag{6.47}$$

where the matrices are

$$G_1=\sum_{k=1}^{\infty}(-1)^k\frac{1}{\mu^k}\left(\Phi_o^{-1}\left(\Psi_1(0)\Psi_1^{\mathrm{T}}(0)\right)^{-1}\right)^k,$$

$$G_2=\sum_{k=0}^{\infty}(-1)^k\frac{1}{\mu^k}\left(\left(\Psi_1(0)\Psi_1^{\mathrm{T}}(0)\right)^{-1}\Phi_o^{-1}\right)^k,$$

$$M_2=\int_0^t e^{\frac{1}{\epsilon}A_1^{\mathrm{T}}\tau}C^{\mathrm{T}}\left(C\sigma\Psi_1\left(\hat{W}-\hat{W}_f\right)+C\varepsilon_1\right)d\tau.$$

Furthermore, the following inequalities can be obtained

$$\|G_1\|\le\sum_{k=1}^{\infty}\frac{\mu^\star}{\mu}=\frac{\mu^\star}{\mu-\mu^\star},\ \|G_2\|\le\sum_{k=0}^{\infty}\frac{\mu^\star}{\mu}=\frac{\mu}{\mu-\mu^\star},\ \|G_3\|\le\frac{1}{\mu}\frac{k_1}{k_2}, \tag{6.48}$$

and

$$G_3\triangleq\frac{1}{\gamma\epsilon^{2n}}\Phi_o^{-1}\left(\Psi_1(0)\Psi_1^{\mathrm{T}}(0)\right)^{-1},\ k_1\triangleq\left\|\left(\Psi_1(0)\Psi_1^{\mathrm{T}}(0)\right)^{-1}\right\|\in\mathbf{R}^+,$$

and there exists a constant $k_2 \in \mathbf{R}^+$ such that $0 < k_2 < \lambda_{\min}(\Phi_o)$ due to its positive definiteness; $\mu^\star \triangleq \dfrac{k_1}{k_2} \in \mathbf{R}^+$ satisfies $\mu \ge \mu^\star$.

Furthermore, through Eqs. (6.39), (6.46) and (6.48), we can obtain

$$\left\|\Psi_1\tilde{W}\right\|\le k_0\frac{\mu^\star}{\mu-\mu^\star}\left\|\Psi_1(0)\tilde{W}(0)\right\|+\frac{k_0^2}{k_2}\frac{\mu}{\mu-\mu^\star}\frac{\epsilon}{4\lambda_0}\left(\sigma M_3+\|\varepsilon_1\|\right), \tag{6.49}$$

where

$$M_3=\sup_{0\le\tau\le t}\|\Psi_1\tilde{W}\|+\sup_{0\le\tau\le t}\|\Psi_1\tilde{W}_f\|.$$

By integrating $\frac{d\Psi_1 W_f}{dt} = \frac{A_1}{\epsilon}\Psi_1 \tilde{W}_f + \tau_f \Psi_1 \left(\hat{W} - \hat{W}_f\right)$, we can obtain

$$\Psi_1 \tilde{W}_f \leq \tau_f \int_0^t e^{\frac{A_1}{\epsilon}(\tau - t)} \Psi_1 \left(\tilde{W} - \tilde{W}_f\right) d\tau + e^{\frac{A_1}{\epsilon}t} \Psi_1 (0) \tilde{W}_f (0). \tag{6.50}$$

Therefore, the following inequality can be obtained

$$\begin{aligned} \left\|\Psi_1 \tilde{W}_f\right\| &\leq \tau_f k_0 \frac{\epsilon}{\lambda_0}\left(e^{-\frac{\lambda_0}{\epsilon}t} - e^{-\frac{2\lambda_0}{\epsilon}t}\right) M_3 + k_0 e^{-\frac{\lambda_0}{\epsilon}t} \|\eta (0)\| \\ &\leq k_0 \|\eta (0)\| + \tau_f k_0 \frac{\epsilon}{4\lambda_0} \sup_{0 \leq \tau \leq t} \|\Psi_1 \tilde{W} + \tau_f k_0 \frac{\epsilon}{4\lambda_0} \sup_{0 \leq \tau \leq t} \|\Psi_1 \tilde{W}_f\|. \end{aligned} \tag{6.51}$$

There exist $\epsilon^\star > 0$ and $\tau_f^\star > 0$ such that for $0 < \tau_f < \tau_f^\star$ and $0 < \epsilon < \epsilon^\star$, we have $\tau_f k_0 \frac{\epsilon}{4\lambda_0} < 1$. Therefore, the following inequality can be obtained

$$\sup_{0 \leq \tau \leq t} \|\Psi_1 \tilde{W}_f\| \leq \frac{\tau_f k_0 \epsilon}{4\lambda_0 - \epsilon k_0 \tau_f} \sup_{0 \leq \tau \leq t} \|\Psi_1 \tilde{W}\| + \frac{4\lambda_0 k_0 \|\eta (0)\|}{4\lambda_0 - \epsilon k_0 \tau_f}. \tag{6.52}$$

There exist $\mu_1^\star > \mu^\star$, $0 < \tau_{f1}^\star < \tau_f^\star$, $\sigma^\star > 0$ and $0 < \epsilon_1^\star < \epsilon^\star$ such that for $\mu > \mu_1^\star, 0 < \tau_f < \tau_{f1}^\star, 0 < \sigma < \sigma^\star$ and $0 < \epsilon < \epsilon_1^\star$, we have $\sigma \frac{k_0^2}{k_2} \frac{\mu}{\mu - \mu^\star} \frac{\epsilon}{4\lambda_0 - \epsilon k_0 \tau_f} < 1$ and $\Delta \triangleq 1 - \sigma \frac{k_0^2}{k_2} \frac{\mu}{\mu - \mu^\star} \frac{\epsilon}{4\lambda_0 - \epsilon k_0 \tau_f} \in \mathbf{R}^+$. According to Eqs. (6.40) and (6.48)–(6.52), we can obtain

$$\begin{aligned} \sup_{0 \leq \tau \leq t} \|\eta\| \leq{}& \frac{k_0}{\Delta} \frac{\mu^\star}{\mu - \mu^\star} \|\eta (0)\| + \frac{\sigma}{\Delta} \frac{k_0^3}{k_2} \frac{\mu}{\mu - \mu^\star} \frac{\epsilon}{4\lambda_0 - \epsilon k_0 \tau_f} \|\eta (0)\| \\ &+ \frac{k_0^2}{\Delta k_2} \frac{\mu}{\mu - \mu^\star} \frac{\epsilon}{4\lambda_0} \varepsilon_d + \varepsilon_d. \end{aligned} \tag{6.53}$$

According to $\epsilon^{2n} \|\eta\|^2 \leq \|\tilde{x}_o\|^2 \leq \|\eta\|^2$, the estimation error satisfies

$$\begin{aligned} \sup_{0 \leq \tau \leq t} \|\tilde{x}_o\| \leq{}& \frac{k_0}{\Delta} \frac{\mu^\star \epsilon^{-n}}{\mu - \mu^\star} \|\tilde{x}_o (0)\| + \frac{\sigma}{\Delta} \frac{k_0^3}{k_2} \frac{\mu}{\mu - \mu^\star} \frac{\epsilon^{1-n}}{4\lambda_0 - \epsilon k_0 \tau_f} \|\tilde{x}_o (0)\| \\ &+ \frac{k_0^2}{\Delta k_2} \frac{\mu}{\mu - \mu^\star} \frac{\epsilon}{4\lambda_0} \varepsilon_d + \varepsilon_d. \end{aligned} \tag{6.54}$$

Based on Eqs. (6.52) and (6.53) as well as $\|\nu\| \leq \bar{\nu}$ and $\|\varepsilon_1\| \leq \varepsilon_{d1}$, we can obtain

$$\sup_{0\le\tau\le t} \|\Psi_1 \tilde{W}\| \le \frac{\sigma}{\Delta}\frac{k_0^3}{k_2}\frac{\mu}{\mu-\mu^\star}\frac{\epsilon\|\eta(0)\|}{4\lambda_0-\epsilon k_0\tau_f} + \frac{k_0^2}{\Delta k_2}\frac{\mu}{\mu-\mu^\star}\frac{\epsilon}{4\lambda_0}\varepsilon_{d1} + \frac{k_0}{\Delta}\frac{\mu^\star\|\eta(0)\|}{\mu-\mu^\star}, \tag{6.55}$$

$$\begin{aligned}\sup_{0\le\tau\le t} \|\Psi_1 \tilde{W}_f\| \le{}& \frac{\tau_f k_0^3\epsilon}{4\lambda_0-\epsilon k_0\tau_f}\frac{\mu}{\mu-\mu^\star}\frac{\sigma k_0}{\Delta k_2}\frac{\epsilon\|\eta(0)\|}{4\lambda_0-\epsilon k_0\tau_f} \\ &+ \frac{\tau_f k_0\epsilon}{4\lambda_0-\epsilon k_0\tau_f}\frac{k_0}{\Delta}\frac{\mu^\star\|\eta(0)\|}{\mu-\mu^\star} \\ &+ \frac{\tau_f k_0\epsilon}{4\lambda_0-\epsilon k_0\tau_f}\frac{k_0^2}{\Delta k_2}\frac{\mu}{\mu-\mu^\star}\frac{\epsilon}{4\lambda_0}\varepsilon_{d1} \\ &+ \frac{4k_0\lambda_0\|\eta(0)\|}{4\lambda_0-\epsilon k_0\tau_f}. \end{aligned} \tag{6.56}$$

Therefore, based on Eqs. (6.55) and (6.56), we can conclude that $\sup_{0\le\tau\le t} \|\Psi_1 \tilde{W}_f\|$ and $\sup_{0\le\tau\le t} \|\Psi_1 \tilde{W}\|$ are bounded. □

Remark 6.3 Note that the peak phenomenon of uncertainty estimation mainly comes from the initial estimation error $\tilde{x}_o$, and it is difficult to determine the initial value of uncertainty in practical applications. It can be clearly seen from Eq. (6.54) that MESO fusion can effectively reduce the peak by adjusting the self-adaptive design parameters ϵ, μ, σ and τ_f. In addition, the transient estimation error of MESO fusion, its peak can reach $O\left(\frac{\|\tilde{x}_o(0)\|}{\mu\epsilon^n} + \frac{\sigma\|\tilde{x}_o(0)\|}{\epsilon^{n-1}}\right)$, which is superior to the transient peak $O\left(\frac{\|\tilde{x}_o\|}{\epsilon^n}\right)$ of ESO in Ref. [10], and is less affected by the approximation error and measurement noise in (6.28). Unlike the methods in Ref. [7] which involve reducing the observer gain and sacrificing the convergence rate, MESO achieves desirable peak suppression through the low-frequency adaptive fusion of multiple ESOs, while maintaining rapid convergence and steady-state accuracy in uncertainty estimation.

Theorem 6.2 *Consider the uncertain nonlinear system (6.23) and the proposed MESO fusion (6.25), (6.26) and low-frequency adaptive parameter learning law (6.31). Assume that conditions in Assumption 4.1 and 4.2 and Theorem 6.1 are satisfied, and the initial conditions of MESO fusion belong to the compact subset of* $\boldsymbol{R}^{n+1}$, *then there exists* $\epsilon^\star > 0$, *such that for* $0 < \epsilon \le \epsilon^\star$, *MESO fusion is ultimately bounded.*

Proof Consider the Lyapunov function

$$V_\eta = \eta^{\mathrm{T}} P \eta. \tag{6.57}$$

Since $\lambda_{\min}(P) I \leq P \leq \lambda_{\max}(P)$, we have

$$\lambda_{\min}(P)\|\eta\|^2 \leq V_\eta \leq \lambda_{\max}(P)\|\eta\|^2. \tag{6.58}$$

The derivative of Eq. (6.57) is written as

$$\begin{aligned}\dot{V}_\eta = & -\frac{1}{\epsilon}\eta^{\mathrm{T}}\eta + 2\sigma\epsilon^{2n}\eta^{\mathrm{T}}P\Psi_1 K\Psi_1^T C\Psi_1\left(\tilde{W}_f - \tilde{W}\right) \\ & - 2\epsilon^{2n}\eta^{\mathrm{T}}P\Psi_1 K\Psi_1^{\mathrm{T}} C^{\mathrm{T}} C\eta + 2\eta^{\mathrm{T}} P D h.\end{aligned} \tag{6.59}$$

Furthermore, according to Eq. (6.57) and $h \leq L_0$, there exist constants a_0, a_1, and a_2 such that

$$\dot{V}_\eta \leq -\left(\frac{1}{\epsilon} - a_0\right)\|\eta\|^2 + \frac{a_1}{\epsilon^{n+1}}\bar{\upsilon}\|\eta\| + a_2\|\eta\|, \tag{6.60}$$

and the positive constants are

$$\begin{aligned}
a_0 &\geq 2\gamma\epsilon^{2n}\|P\|k_0^2 e^{-\frac{2\lambda_0 t}{\epsilon}}\|\Psi_1(0)\|^2,\\
a_1 &\geq 2\|H_1\|\|P\| + \epsilon L_2 + \sigma L_1\frac{k_0^2}{\Delta k_2}\frac{\mu}{\mu-\mu^\star}\frac{\epsilon^2}{4\lambda_0} + \sigma L_1\frac{\tau_f k_0\epsilon}{4\lambda_0 - \epsilon k_0\tau_f}\frac{k_0^2}{\Delta k_2}\frac{\mu}{\mu-\mu^\star}\frac{\epsilon^2}{4\lambda_0},\\
a_2 &\geq 2L_0\|P\| + \sigma L_1\frac{k_0}{\Delta}\frac{\mu^\star\|\eta(0)\|}{\mu-\mu^\star} + \sigma L_1\frac{4k_0\lambda_0\|\eta(0)\|}{4\lambda_0 - \epsilon k_0\tau_f} + \sigma L_1\frac{k_0^2}{\Delta k_2}\frac{\mu}{\mu-\mu^\star}\frac{\epsilon}{4\lambda_0}\varepsilon_{d1}\\
&\quad + \sigma L_1\frac{\tau_f k_0^3\epsilon}{4\lambda_0 - \epsilon k_0\tau_f}\frac{\mu}{\mu-\mu^\star}\frac{\sigma k_0}{\Delta k_2}\frac{\epsilon\|\eta(0)\|}{4\lambda_0 - \epsilon k_0\tau_f} + \sigma L_1\frac{\tau_f k_0\epsilon}{4\lambda_0 - \epsilon k_0\tau_f}\frac{k_0}{\Delta}\frac{\mu^\star\|\eta(0)\|}{\mu-\mu^\star}\\
&\quad + \sigma L_1\frac{\tau_f k_0\epsilon}{4\lambda_0 - \epsilon k_0\tau_f}\frac{k_0^2}{\Delta k_2}\frac{\mu}{\mu-\mu^\star}\frac{\epsilon}{4\lambda_0}\varepsilon_{d1} + \sigma L_1\frac{\sigma}{\Delta}\frac{k_0^3}{k_2}\frac{\mu}{\mu-\mu^\star}\frac{\epsilon\|\eta(0)\|}{4\lambda_0 - \epsilon k_0\tau_f},\\
L_1 &\triangleq 2\gamma\epsilon^{2n}\|P\|k_0^2 e^{-\frac{2\lambda_0 t}{\epsilon}}\|\Psi_1(0)\|^2.
\end{aligned}$$

There exists $\epsilon^\star > 0$, by choosing $0 < \epsilon \leq \epsilon^\star$, we can get $a_0 \leq \frac{1}{2\epsilon}$, the result is

$$\dot{V}_\eta \leq -\frac{1}{2\epsilon}\|\eta\|^2 + \frac{a_1}{\epsilon^{n+1}}\bar{\upsilon}\|\eta\| + a_2\|\eta\|.$$

Through Eq. (6.58), we have

$$\dot{V}_\eta \leq -\frac{V_\eta}{2\epsilon\lambda_{\max}(P)} + \frac{a_1\sqrt{V_\eta}}{\epsilon^{n+1}\sqrt{\lambda_{\min}(P)}}\bar{\upsilon} + a_2\frac{\sqrt{V_\eta}}{\lambda_{\min}(P)}. \tag{6.61}$$

Then, $W_\eta \triangleq \sqrt{V_\eta}$. For $V_\eta \neq 0$, the derivative of W_η satisfies

$$\dot{W}_\eta \leq -\frac{W_\eta}{4\epsilon\lambda_{\max}(P)} + \frac{a_1\bar{\nu}}{2\epsilon^{n+1}\sqrt{\lambda_{\min}(P)}} + \frac{a_2}{2\sqrt{\lambda_{\min}(P)}}. \tag{6.62}$$

If $V_\eta \neq 0$, $D^+W_\eta = \dot{W}_\eta$. In order to prove that D^+W_η satisfies Eq. (6.62), when $V_\eta = 0$, we need to prove that D^+W_η satisfies

$$D^+W_\eta \leq \frac{a_1\bar{\nu}}{2\epsilon^{n+1}\sqrt{\lambda_{\min}(P)}} + \frac{a_2}{2\sqrt{\lambda_{\min}(P)}}. \tag{6.63}$$

According to the definition of Dini derivative $D^+W = \lim\sup_{h\to 0^+} \frac{W(t+h)-W(t)}{h}$ and $\eta = 0\left(V_\eta = 0,\ W_\eta = 0\right)$, we can get

$$\begin{aligned} D^+W_\eta &\leq \lim\sup_{h\to 0^+} \frac{\sqrt{\lambda_{\max}(P)}}{h}\|\eta(t+h)\| \\ &\leq \frac{1}{2\sqrt{\lambda_{\min}(P)}}\left(\frac{a_1}{\epsilon^{n+1}}\bar{\nu} + a_2\right). \end{aligned} \tag{6.64}$$

Therefore, it can be seen that Eq. (6.62) is still valid. Define a bounded set $\Omega_\eta \triangleq \{\eta \in \mathbf{R}^{n+1} | W_\eta \leq 4\epsilon\lambda(P)\omega_\eta\}$, where $\omega_\eta = \frac{1}{2\sqrt{\lambda_{\min}(P)}}\left(\frac{a_1}{\epsilon^{n+1}}\bar{\nu} + a_2\right)$. When $\eta \in \{\mathbf{R}^{n+1} \backslash \Omega_\eta\}$, $\dot{W}_\eta \leq 0$. According to Eqs. (6.58) and (6.62) and the theorem about boundedness and stability in [121], it can be concluded that MESO is ultimately bounded. □

6.5.2 Dynamics Wheel Velocity Disturbance Rejection Control and Stability Analysis

The active docking unit adopts the same unit dynamics as Chap. 3, with the control objective being to track the desired linear velocity and angular velocity converted into wheel velocity commands ω_{dir} and ω_{dil}. Due to the saturation of velocity and acceleration, the dynamics reference wheel velocity and its derivative are bounded.

The feedback controller for the wheel velocity dynamics of the active docking unit is also designed as

$$\begin{aligned} \tau_{ir} &= D_{11}\left(\dot{\omega}_{dir} - k_{i\omega}\tilde{\omega}_{ir}\right) + D_{12}\dot{\omega}_{il} + f_c\left(\omega_{ir}\right) + D_{11}\hat{d}_{ir}, \\ \tau_{il} &= D_{21}\left(\dot{\omega}_{dil} - k_{i\omega}\tilde{\omega}_{il}\right) + D_{22}\dot{\omega}_{ir} + f_c\left(\omega_{il}\right) + D_{22}\hat{d}_{il}, \end{aligned} \tag{6.65}$$

where $k_{i\omega} \in \mathbf{R}_+$ is the control gain, $\tilde{\omega}_{ir} = \omega_{dir} - \omega_{ir}$ and $\tilde{\omega}_{il} = \omega_{dil} - \omega_{il}$ are the wheel velocity tracking errors; $\hat{d}_{ir}, \hat{d}_{il}, \tilde{d}_{ir}$ and $\tilde{d}_{il}$ are the total disturbance estimation signals and their errors estimated by MESO adaptive fusion.

Taking the left wheel dynamics control model (4.24) as an example, according to theorem (6.2), three ESO adaptive fusion disturbance estimation schemes are designed, and the specific form of the jth ESO is as follows

$$\begin{aligned} \dot{\hat{\omega}}_{ijl} &= \frac{1}{D_{11}}\tau_{il} + \hat{d}_{ijl} + \beta_{i1}\left(\omega_{il} - \hat{\omega}_{il}\right), \\ \dot{\hat{d}}_{ijl} &= \beta_{i2}\left(\omega_{il} - \hat{\omega}_{ijl}\right), \end{aligned} \tag{6.66}$$

where $\hat{\omega}_{ijl}$ and $\hat{d}_{ijl}$ represent the wheel velocity and the total disturbance estimation of the jth ESO, and β_{i1} and β_{i2} are the ESO design gains. The multi-ESO adopts the adaptive fusion scheme shown in Eqs. (6.26) and (6.31). $\hat{d}_{il}$ and $\hat{d}_{ir}$ are the extended state estimates in $\hat{x}_o$. According to theorem (6.1), the upper bounds $\bar{d}_{il}$ and $\bar{d}_{ir}$ of the disturbance estimation errors $\tilde{d}_{il}$ and $\tilde{d}_{ir}$ are bounded functions.

The Lyapunov function of wheel velocity dynamics is chosen as

$$V_{i\omega} = \frac{1}{2}\tilde{\omega}_{il}^2 + \frac{1}{2}\tilde{\omega}_{ir}^2. \tag{6.67}$$

Its derivative can be easily obtained as

$$\begin{aligned} \dot{V}_{i\omega} &= -k_{i\omega}\tilde{\omega}_{il}^2 + \tilde{\omega}_{il}\tilde{d}_{il} - k_{i\omega}\tilde{\omega}_{ir}^2 + \tilde{\omega}_{ir}\tilde{d}_{ir} \\ &\le -k_{i\omega}\tilde{\omega}_{il}^2 + \frac{1}{4}\tilde{\omega}_{il}^2 - k_{i\omega}\tilde{\omega}_{ir}^2 + \frac{1}{4}\tilde{\omega}_{ir}^2 + \bar{d}_{il}^2 + \bar{d}_{ir}^2 \\ &\le -\bar{k}_{iw}V_{i\omega} + \Delta_{i\omega}, \end{aligned} \tag{6.68}$$

where $\bar{k}_{iw} = 2\left(k_{i\omega} - \frac{1}{4}\right) \in \mathbf{R}_{+}$, $\Delta_{i\omega} = \bar{d}_{il}^2 + \bar{d}_{ir}^2$.

The wheel velocity error can actively control the tracking errors of the linear velocity and angular velocity of the interface unit

$$v_{ie} = \frac{r_i\left(\tilde{\omega}_{ir} + \tilde{\omega}_{il}\right)}{2}, \; w_{ie} = \frac{r_i\left(\tilde{\omega}_{ir} - \tilde{\omega}_{il}\right)}{2b_i}. \tag{6.69}$$

Remark 6.4 From Eqs. (6.67) and (6.68), it can be seen that by adjusting the dynamic feedback control cabinet and disturbance estimation gain, $\tilde{\omega}_{il}$ and $\tilde{\omega}_{ir}$ will eventually converge to a sufficiently small region. The velocity error of the lower layer dynamics will be equivalent to the uncertainty transmitted to the kinematic level. In addition to adjusting the kinematic control gain, the most appropriate way is to adjust the control gain of the lower layer, so that the velocity tracking relative to the kinematic level has faster convergence characteristics and high-precision steady-state performance, in order to achieve the desired kinematic trajectory control performance.

6.6 Safe Set Closure Control Under Nested Motion Saturation

Considering the saturation of the nested motion of the reconstructable swarm unit velocity command, a hierarchical safety control framework for adaptive anti-saturation trajectory tracking based on the complete kinematic model of the upper layer obstacle avoidance guidance module and the non-integral constraint model of the lower layer is designed, as shown in Fig. 6.4, to complete the trajectory control task of the remote assembly stage.

6.6.1 Upper Layer Integral Model Adaptive Obstacle Avoidance Guidance Module

For the sake of design simplicity and task feasibility, the upper layer guidance module adopts the following virtual integral kinematic model

$$\dot{\varphi}_1 = \varphi_2, \quad \dot{\varphi}_2 = a^{\varphi}, \tag{6.70}$$

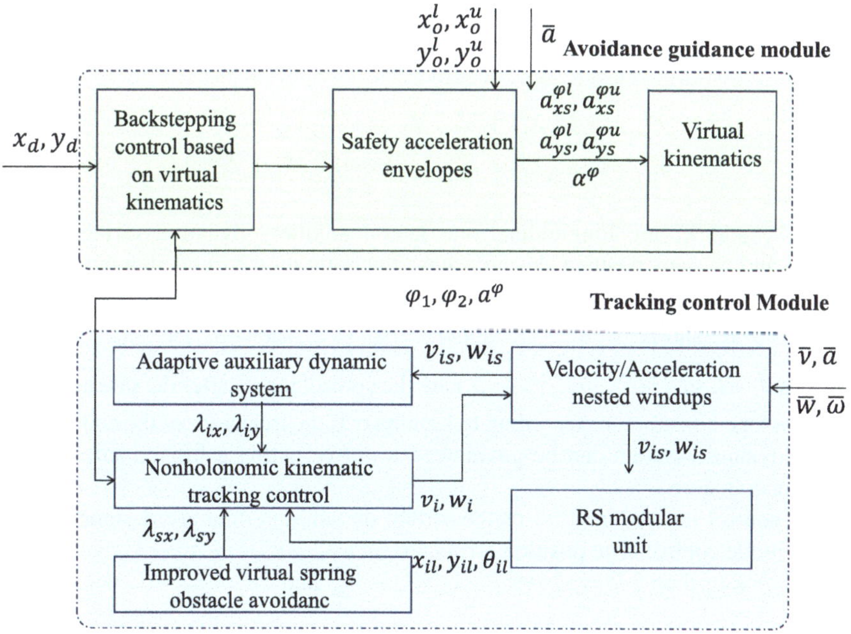

Fig. 6.4 Remote guidance—Follow hierarchical security constraint control framework

where, $\varphi_1 = [x^\varphi, y^\varphi]^{\mathrm{T}} \in \mathbf{R}^2$ and $\varphi_2 = \left[v_x^\varphi, v_y^\varphi\right]^{\mathrm{T}} \in \mathbf{R}^2$ are the position and velocity states of the virtual kinematic dynamics; $a^\varphi = \left[a_x^\varphi, a_y^\varphi\right]^{\mathrm{T}} \in \mathbf{R}^2$ is the virtual dynamics acceleration, with initial values satisfying $\varphi_1(0) = p_{ci}(0)$ and $\varphi_2(0) = \dot{p}_{ci}(0)$; x^φ, y^φ, v_x^φ, v_y^φ, a_x^φ and a_y^φ are the longitudinal and lateral positions, velocities and accelerations of the virtual dynamic system in the inertial system, respectively.

Referring to the safety barrier mechanism [6], the design of the longitudinal and lateral safety acceleration envelopes near obstacles during the approach of the remote stage is as follows

$$\begin{aligned} a_{xs}^{\varphi t} &= -\alpha \mathcal{B}\left(\bar{x}_t^\varphi, \dot{\bar{x}}_t^\varphi\right) + \mathcal{H}\left(\bar{x}_t^\varphi\right)\dot{\bar{x}}_t^\varphi, \\ a_{ys}^{\varphi t} &= -\alpha \mathcal{B}\left(\bar{y}_t^\varphi, \dot{\bar{y}}_t^\varphi\right) + \mathcal{H}\left(\bar{y}_t^\varphi\right)\dot{\bar{y}}_t^\varphi, \quad t = u, l, \end{aligned} \tag{6.71}$$

where, $\bar{x}_u^\varphi = x^\varphi - x_o^u$, $\bar{x}_l^\varphi = x_o^l - x^\varphi$, $\bar{y}_u^\varphi = y^\varphi - y_o^u$ and $\bar{y}_l^\varphi = y_o^l - y^\varphi$, and x_o^u, x_o^l, y_o^u and y_o^l are the safety barriers set according to the obstacle coordinates (x_o, y_o). The designed control inputs for the virtual dynamic system, u_x^φ and u_y^φ, satisfy the following saturation constraints

$$a_x^\varphi = \mathrm{sat}\left(-\bar{\alpha}, a_x^\varphi, \bar{\alpha}\right), a_y^\varphi = \mathrm{sat}\left(-\bar{\alpha}, a_y^\varphi, \bar{\alpha}\right), \tag{6.72}$$

where $\bar{\alpha} < \bar{a} \in \mathbf{R}_+$.

Acceleration envelope constraints can also trigger saturation oscillations. The following adaptive auxiliary dynamic system is introduced to smooth the virtual dynamic trajectory

$$\dot{\lambda}_{1x}^\varphi = -k_{1x}^\varphi \lambda_{1x}^\varphi + \lambda_{2x}^\varphi, \dot{\lambda}_{2x}^\varphi = -k_{2x}^\varphi \lambda_{2x}^\varphi + \Delta_x^\varphi, \tag{6.73}$$

$$\dot{\lambda}_{1y}^\varphi = -k_{1y}^\varphi \lambda_{1y}^\varphi + \lambda_{2y}^\varphi, \dot{\lambda}_{2y}^\varphi = -k_{2y}^\varphi \lambda_{2y}^\varphi + \Delta_y^\varphi, \tag{6.74}$$

where λ_x^φ and λ_y^φ are longitudinal and lateral auxiliary dynamic variables; k_x^φ, k_y^φ, γ_x^φ and γ_y^φ are positive design gains; the saturation virtual dynamic acceleration and the to-be-designed longitudinal and lateral control input deviations $\Delta_x^\varphi = a_x^\varphi - u_x^\varphi$ and $\Delta_y^\varphi = a_y^\varphi - u_y^\varphi$, and $a_x^\varphi = \mathrm{sat}\left(-\bar{\alpha}, \mathrm{sat}\left(-a_{xs}^{\varphi l}, u_x^\varphi, a_{xs}^{\varphi u}\right), \bar{\alpha}\right)$ and $a_y^\varphi = \mathrm{sat}\left(-\bar{\alpha}, \mathrm{sat}\left(-a_{ys}^{\varphi l}, u_y^\varphi, a_{ys}^{\varphi u}\right), \bar{\alpha}\right)$ are the control inputs after the safe acceleration envelope constraints. According to Lemma 6.1, the trajectory of the constrained virtual dynamic system can be guaranteed to move within a limited safe region, avoiding collisions.

The control laws u_x^φ and u_y^φ of the virtual dynamic system are designed using sliding mode control. The position errors are set as $z_{1x} = x^\varphi - x_d - \lambda_{1x}$ and $z_{1y} = y^\varphi - y_d - \lambda_{1y}$, then $\dot{z}_{1x} = v_x^\varphi - \dot{x}_d + k_{1x}\lambda_{1x}^\varphi - \lambda_{2x}^\varphi$ and $\dot{z}_{1y} = v_y^\varphi - \dot{y}_d + k_{1y}\lambda_{1x}^\varphi - \lambda_{2y}^\varphi$.

The sliding mode functions are set as $s_x^\varphi = c_1 z_{1x} + \dot{z}_{1x}$ and $s_y^\varphi = c_1 z_{1y} + \dot{z}_{1y}$, where $c_1 > 0$ is the control gain, so we have

$$\begin{aligned}
\dot{s}_x^\varphi &= c_1\dot{z}_{1x} + \ddot{z}_{1x} = c_1\dot{z}_{1x} + a_x^\varphi - \ddot{x}_d - \ddot{\lambda}_{1x} \\
&= c_1\dot{z}_{1x} + a_x^\varphi - \ddot{x}_d - \left(-k_{1x}^\varphi\dot{\lambda}_{1x}^\varphi + \dot{\lambda}_{2x}^\varphi\right) \\
&= c_1\dot{z}_{1x} + u_x^\varphi + \Delta_x^\varphi - \ddot{x}_d + k_{1x}^\varphi\left(-k_{1x}^\varphi\lambda_{1x}^\varphi + \lambda_{2x}^\varphi\right) + k_{2x}^\varphi\lambda_{2x}^\varphi - \Delta_x^\varphi \\
&= c_1\dot{z}_{1x} + u_x^\varphi - \ddot{x}_d + k_{1x}^\varphi\left(-k_{1x}^\varphi\lambda_{1x}^\varphi + \lambda_{2x}^\varphi\right) + k_{2x}^\varphi\lambda_{2x}^\varphi, \qquad (6.75)\\
\dot{s}_y^\varphi &= c_1\dot{z}_{1y} + \ddot{z}_{1y} = c_1\dot{z}_{1y} + a_y^\varphi - \ddot{y}_d - \ddot{\lambda}_{1y} \\
&= c_1\dot{z}_{1y} + a_y^\varphi - \ddot{y}_d - \left(-k_{1y}^\varphi\dot{\lambda}_{1y}^\varphi + \dot{\lambda}_{2y}^\varphi\right) \\
&= c_1\dot{z}_{1y} + u_y^\varphi + \Delta_y^\varphi - \ddot{y}_d + k_{1y}^\varphi\left(-k_{1y}^\varphi\lambda_{1y}^\varphi + \lambda_{2y}^\varphi\right) + k_{2y}^\varphi\lambda_{2y}^\varphi - \Delta_y^\varphi \\
&= c_1\dot{z}_{1y} + u_y^\varphi - \ddot{y}_d + k_{1y}^\varphi\left(-k_{1y}^\varphi\lambda_{1y}^\varphi + \lambda_{2y}^\varphi\right) + k_{2y}^\varphi\lambda_{2y}^\varphi. \qquad (6.76)
\end{aligned}$$

The designed upper-layer obstacle avoidance guidance controllers for longitudinal and lateral directions are

$$u_x^\varphi = -c_1\dot{z}_{1x} + \ddot{x}_d - k_{1x}^\varphi\left(-k_{1x}^\varphi\lambda_{1x}^\varphi + \lambda_{2x}^\varphi\right) - k_{2x}^\varphi\lambda_{2x}^\varphi - \eta_x^\varphi \operatorname{sgn}\left(s_x^\varphi\right), \qquad (6.77)$$

$$u_y^\varphi = -c_1\dot{z}_{1y} + \ddot{y}_d - k_{1y}^\varphi\left(-k_{1y}^\varphi\lambda_{1y}^\varphi + \lambda_{2y}^\varphi\right) - k_{2y}^\varphi\lambda_{2y}^\varphi - \eta_y^\varphi \operatorname{sgn}\left(s_y^\varphi\right), \qquad (6.78)$$

where η_x^φ and $\eta_y^\varphi \in \mathbf{R}_+$.

Therefore, we define the Lyapunov function as

$$V_{1\varphi} = \frac{1}{2}\left(s_x^\varphi\right)^2 + \frac{1}{2}\left(s_y^\varphi\right)^2, \qquad (6.79)$$

Combining the control laws (6.77)–(6.78) and the sliding mode dynamics (6.75)–(6.76), the derivative of Eq. (6.79) satisfies

$$\dot{V}_{1\varphi} = -c_1\left(s_x^\varphi\right)^2 - c_1\left(s_y^\varphi\right)^2 - \eta_x^\varphi|s_x^\varphi| - \eta_y^\varphi|s_y^\varphi| \le c_\varphi V_{1\varphi}, \qquad (6.80)$$

where $c_\varphi = 2c_1 \in \mathbf{R}_+$. In summary, $V_{1\varphi}$ is positive definite, $\dot{V}_{1\varphi}$ is negative definite, therefore the designed virtual system control laws (6.77) and (6.78) can make z_{1x}, z_{2x}, z_{1y}, and z_{2y} asymptotically converge to 0. Meanwhile, according to Lemma 6.2, the virtual dynamic system can always ensure the position of the active docking unit in the virtual dynamic system to be in the safe region under the constraint of safe acceleration envelope.

Remark 6.5 The upper layer constrained a^φ can generate the lower layer tracking expected signals φ_1 and φ_2 through double and single integrals, so the upper layer guidance module is equivalent to the model uncertainty of the lower layer tracking in terms of the actual expected tracking error. $x^\varphi - x_d$ and $y^\varphi - y_d$ are the upper layer position tracking errors, and $v_x^\varphi - \dot{x}_d$ and $v_y^\varphi - \dot{y}_d$ are the upper layer velocity tracking errors. According to Eqs. (6.79) and (6.80), it is easy to obtain that the upper bounds $\delta_{x\varphi}$, $\delta_{y\varphi}$, $\delta_{vx\varphi}$, and $\delta_{vy\varphi}$ of $|x^{\varphi} - x_d|$, $|y^{\varphi} - y_d|$, $|v_x^{\varphi} - \dot{x}_d|$, and $|v_y^{\varphi} - \dot{y}_d|$ are bounded, and the size of the boundaries can be adjusted through gain regulation.

6.6.2 *Lower Layer Kinematic Model Adaptive Anti-saturation Trajectory Controller*

By integrating the adaptive anti-saturation and improved spring obstacle avoidance strategy, a dual-layer constraint mechanism is formed with the upper layer guidance, and a trajectory tracking control scheme is designed under the velocity/acceleration nested saturation constraint.

Combining the kinematic model (6.1) and nested saturation (6.3), the auxiliary dynamic system of the active docking unit with adaptive anti-saturation is designed as

$$\begin{aligned} \dot{\lambda}_{ix} &= \cosh^2(\lambda_{ix})\left(-k_{ix}\lambda_{ix} + \Delta_{ix}\right)/\gamma_{ix}, \\ \dot{\lambda}_{iy} &= \cosh^2(\lambda_{iy})\left(-k_{iy}\lambda_{iy} + \Delta_{iy}\right)/\gamma_{iy}, \end{aligned} \tag{6.81}$$

where the saturation velocity and velocity command are distributed in the motion space as

$$\begin{aligned} \Delta_{ix} &= (v_{is} - v_i)\cos(\theta_i) - l_i(w_{is} - w_i)\sin(\theta_i), \\ \Delta_{iy} &= (v_{is} - v_i)\sin(\theta_i) + l_i(w_{is} - w_i)\cos(\theta_i), \end{aligned}$$

λ_{ix} and λ_{iy} are the auxiliary dynamic variables in the longitudinal and lateral directions; k_{ix}, k_{iy}, γ_{ix}, and γ_{iy} are positive adaptive gains.

The force exerted by the obstacle (x_o, y_o) on the active docking unit is expressed as

$$F_{ro} = \begin{cases} \left|k_r \tanh\left(\tan\left(\frac{\pi}{2}\frac{\rho_{sR}-\rho_{ro}}{\rho_{sR}-\rho_{sr}}\right)\right)\right|, & \rho_{ro} \le \rho_{sR}, \\ 0, & \rho_{ro} > \rho_{sR}, \end{cases} \tag{6.82}$$

Here, $\rho_{ro} = \| [x_{il}, y_{il}]^{\mathrm{T}} - [x_o, y_o]^{\mathrm{T}} \|$ is the distance between the active contact unit and the obstacle, ρ_{sr} and ρ_{sR} are the upper and lower bounds of the obstacle avoidance distance, and k_r is the elasticity coefficient. The design of $k_r \tanh(\cdot)$ is related to the boundedness of energy and the saturation threshold of acceleration. The entire repulsive force function reflects the closed-loop dynamic regulation mechanism of the distance between the active contact unit and the obstacle, and can also introduce an auxiliary dynamic system embedded in the controller to ensure system safety. The specific design is as follows

$$\dot{\lambda}_{sx} = \cosh^2(\lambda_{sx})(-k_{sx}\lambda_x + \Delta_{sx})/\gamma_{sx},$$
$$\dot{\lambda}_{sy} = \cosh^2(\lambda_{sy})(-k_{sy}\lambda_y + \Delta_{sy})/\gamma_{sy}, \tag{6.83}$$

where λ_{sx} and λ_{sy} are auxiliary dynamic variables for correcting velocity, k_{sx}, k_{sy}, γ_{sx} and γ_{sy} are positive self-adaptive gains. Δ_{sx} and Δ_{sy} are designed according to the torque distribution method proposed by Yao Chunshi et al. as

$$\Delta_{sx} = F_{ro}\cos(\theta_{ro}),\ \Delta_{sy} = F_{ro}\sin(\theta_{ro}),$$

where θ_{ro} is the angle between the line of sight of the repulsive force in the coordinate system of the active contact unit and the axis of the cart.

Combining the control model (6.1), the self-adaptive anti-saturation strategy (6.81) and the lower layer obstacle avoidance (6.82), (6.83), the lower layer tracking errors e_{ix}, e_{iy} and control laws u_{ix}, u_{iy} are designed as

$$e_{ix} = x_{il} - x_{\varphi} - \gamma_{ix}\tanh(\lambda_{ix}),$$
$$e_{iy} = y_{il} - y_{\varphi} - \gamma_{iy}\tanh(\lambda_{iy}), \tag{6.84}$$
$$u_{ix} = -c_2 e_{ix} + v_x^{\varphi} - k_{ix}\lambda_{ix} - \gamma_{sx}\tanh(\lambda_{sx}),$$
$$u_{iy} = -c_2 e_{iy} + v_y^{\varphi} - k_{iy}\lambda_{iy} - \gamma_{sy}\tanh(\lambda_{sy}), \tag{6.85}$$

where $c_2 > 0$ is the control gain. Through the matching mapping relationship between linear velocity, angular velocity and longitudinal/lateral direction (6.1), the linear velocity and angular velocity v_i and w_i can be obtained.

Let the overall tracking error in the far distance stage be $\bar{e}_{ix} = x_{il} - x_d - \gamma_{ix}\tanh(\lambda_{ix})$ and $\bar{e}_{iy} = y_{il} - y_d - \gamma_{iy}\tanh(\lambda_{iy})$. To prove the overall stability of the far distance stage, the following Lyapunov function is selected

$$V_{3i} = \frac{1}{2}\bar{e}_{ix}^2 + \frac{1}{2}\bar{e}_{iy}^2. \tag{6.86}$$

Taking its derivative, we obtain

$$\begin{aligned}
\dot{V}_{2i} &= \bar{e}_{ix}\dot{\bar{e}}_{ix} + \bar{e}_{iy}\dot{\bar{e}}_{iy} \\
&= \bar{e}_{ix}(\dot{x}_{il} - \dot{x}_d + k_{ix}\lambda_{ix} - \Delta_{ix}) + \bar{e}_{iy}(\dot{y}_{il} - \dot{y}_d + k_{iy}\lambda_{iy} - \Delta_{iy}) \\
&= \bar{e}_{ix}(u_{ix} + \Delta_{ix} - \dot{x}_d + k_{ix}\lambda_{ix} - \Delta_{ix}) + \bar{e}_{iy}(u_{iy} + \Delta_{iy} - \dot{y}_d + k_{iy}\lambda_{iy} - \Delta_{iy}) \\
&= \bar{e}_{ix}(u_{ix} - \dot{x}_d + k_{ix}\lambda_{ix}) + \bar{e}_{iy}(u_{iy} - \dot{y}_d + k_{iy}\lambda_{iy}).
\end{aligned} \tag{6.87}$$

Further substituting the control law (6.85) into Eq. (6.87), we obtain

$$\begin{aligned}
\dot{V}_{2i} &= \bar{e}_{ix}\left(-c_2\bar{e}_{ix} + c_2\left(x_d - x^{\varphi}\right) + v_x^{\varphi} - k_{ix}\lambda_{ix} - \gamma_{sx}\tanh\left(\lambda_{sx}\right) - \dot{x}_d + k_{ix}\lambda_{ix}\right)\\
&\quad + \bar{e}_{iy}\left(-c_2\bar{e}_{iy} + c_2\left(y_d - y^{\varphi}\right) + v_y^{\varphi} - k_{iy}\lambda_{iy} - \gamma_{sy}\tanh\left(\lambda_{sy}\right) - \dot{y}_d + k_{iy}\lambda_{iy}\right)\\
&= \bar{e}_{ix}\left(-c_2\bar{e}_{ix} + c_2\left(x_d - x^{\varphi}\right) - \gamma_{sx}\tanh\left(\lambda_{sx}\right) + v_x^{\varphi} - \dot{x}_d\right)\\
&\quad + \bar{e}_{iy}\left(-c_2\bar{e}_{iy} + c_2\left(y_d - y^{\varphi}\right) - \gamma_{sy}\tanh\left(\lambda_{sy}\right) + v_y^{\varphi} - \dot{y}_d\right)\\
&= -c_2\bar{e}_{ix}^2 + c_2\bar{e}_{ix}\left(x_d - x^{\varphi}\right) + \bar{e}_{ix}\left(v_x^{\varphi} - \dot{x}_d\right) + \bar{e}_{ix}\gamma_{sx}\tanh\left(\lambda_{sx}\right)\\
&\quad - c_2\bar{e}_{iy}^2 + c_2\bar{e}_{iy}\left(y_d - y^{\varphi}\right) + \bar{e}_{iy}\left(v_y^{\varphi} - \dot{y}_d\right) + \bar{e}_{iy}\gamma_{sy}\tanh\left(\lambda_{sy}\right)\\
&\le -c_2e_{ix}^2 - c_2e_{iy}^2 + \frac{1}{2}e_{ix}^2 + \frac{1}{2}e_{iy}^2 + \frac{1}{2}c_2e_{ix}^2 + \frac{1}{2}c_2e_{iy}^2 + \frac{1}{2}c_2\bar{\delta}_{x\varphi}^2 + \frac{1}{2}c_2\bar{\delta}_{y\varphi}^2\\
&\quad + \bar{\delta}_{vx\varphi}^2 + \bar{\delta}_{vy\varphi}^2 + \gamma_{sx}^2\tanh^2\left(\lambda_{sx}\right) + \gamma_{sy}^2\tanh^2\left(\lambda_{sy}\right)\\
&\le -\bar{c}_2V_{3i} + \Delta_S,
\end{aligned} \tag{6.88}$$

where $\Delta_S > \gamma_{sx}^2\tanh^2\left(\lambda_{sx}\right) + \gamma_{sy}^2\tanh^2\left(\lambda_{sy}\right) + \frac{1}{2}c_2\bar{\delta}_{x\varphi}^2 + \frac{1}{2}c_2\bar{\delta}_{y\varphi}^2 + \bar{\delta}_{vx\varphi}^2 + \bar{\delta}_{vy\varphi}^2 \in \mathbf{R}_+$, $\bar{c}_2 > 2\left(c_2 - 1\right) \in \mathbf{R}_+$.

Combining Eq. (6.88), it can be seen that the final error boundaries of $\bar{e}_{ix}$ and $\bar{e}_{iy}$ are influenced by the control gain and the bounded term Δ_S. According to Lemma 6.2, the bounded signals $\gamma_{sx}^2\tanh^2\left(\lambda_{sx}\right)$ and $\gamma_{sy}^2\tanh^2\left(\lambda_{sy}\right)$ in Δ_S will quickly converge to 0 after the obstacle avoidance is completed. The term $\frac{1}{2}c_2\bar{\delta}_{x\varphi}^2 + \frac{1}{2}c_2\bar{\delta}_{y\varphi}^2 + \bar{\delta}_{vx\varphi}^2 + \bar{\delta}_{vy\varphi}^2$ is related to the upper bound of the control error of the upper-layer virtual dynamic system. By adjusting the control gain of the upper layer, it can quickly converge to 0. Combined with the adjustment of the control gain of the lower-layer tracking module, the overall desired tracking performance can be achieved.

6.7 Fast Convergence Control Within Line-of-Sight Constraints

Considering the active docking unit's perception of the line-of-sight range and the limited near-distance journey during the near-distance stage, we propose a fast convergence control framework within the line-of-sight as shown in Fig. 6.5. Based on the line-of-sight cone and auxiliary dynamic system, we design an adaptive guidance reference trajectory correction strategy, and introduce a composite nonlinear feedback of acceleration instant convergence in the design of the adaptive auxiliary dynamic system and trajectory controller. This allows the active docking unit to quickly converge to the guidance trajectory within the line-of-sight range, accurately deliver the lock to the positioning hole, and complete the lock reconfigurable of different units.

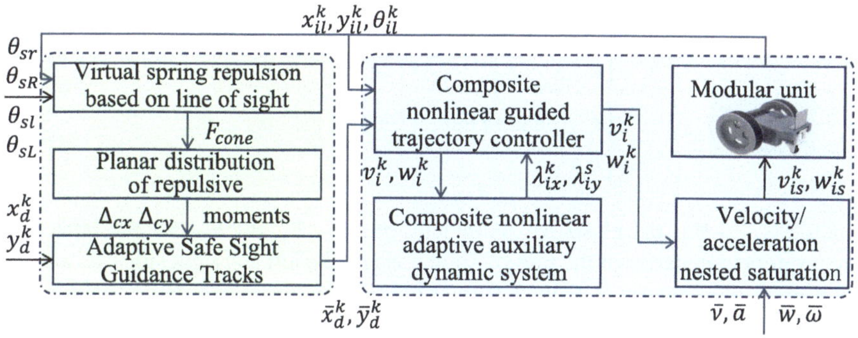

Fig. 6.5 Description and framework of fast convergence control within the safe range of near sight

6.7.1 *Adaptive Compensation Term for Desired Trajectory Based on Line-of-Sight Cone*

The line-of-sight constraint is similar to the environmental obstacle constraint, both of which are safety indicators that cannot be ignored in the control process. At the kinematic level, the near-distance docking control adopts the same kinematic control framework as the far-distance assembly stage, designs longitudinal and lateral motion commands, obtains linear velocity and angular velocity through velocity distribution, but does not directly involve posture adjustment to meet the safety requirements of the line-of-sight constraint. Therefore, based on maintaining the safe line-of-sight cone, the adaptive compensation term of the desired trajectory is designed to adjust the desired trajectory online, promoting the docking target to always be within the sensing line-of-sight range of the active docking unit modality. The idea of adaptive compensation for the desired trajectory is similar to the APF adjustment used by Huang et al. [13]. However, considering the intense reaction of APF when the mobile unit is near the line-of-sight cone, it is easy to trigger saturation, which cannot guarantee the safety of the line-of-sight in actual experiments. Therefore, the spring avoidance strategy in the previous section is used here to design adaptive compensation for Δ_{x_d} and Δ_{y_d}, and the new reference trajectory is

$$\bar{x}_d^k = x_d^k + \Delta_{x_d}, \bar{y}_d^k = \Delta_{y_d}. \tag{6.89}$$

The force amplitude at the measurement position of the active docking unit modality within and outside the line-of-sight cone boundary can be expressed as

$$F_{cone}^{lR} = \begin{cases} k_{cone} | \tanh\left(\tan\left(\frac{\pi}{2}\frac{\theta_{sR}-\theta_{lt}}{\theta_{sR}-\theta_{sr}}\right)\right)|, & \theta_{lt} \le \theta_{sr}, \\ 0, & \theta_{lt} > \theta_{sr}, \end{cases} \tag{6.90}$$

$$F_{cone}^{rR} = \begin{cases} k_{cone} | \tanh\left(\tan\left(\frac{\pi}{2}\frac{\theta_{sR}-\theta_{rt}}{\theta_{sR}-\theta_{sr}}\right)\right)|, & \theta_{rt} \le \theta_{sr}, \\ 0, & \theta_{rt} > \theta_{sR}, \end{cases} \tag{6.91}$$

$$F_{cone}^{lL} = \begin{cases} k_{cone}|\tanh\left(\tan\left(\frac{\pi}{2}\frac{\theta_{sL}-\theta_{lt}}{\theta_{sL}-\theta_{sl}}\right)\right)|, & \theta_{lt} \geq \theta_{sl}, \\ 0, & \theta_{lt} < \theta_{sl}, \end{cases} \tag{6.92}$$

$$F_{cone}^{rL} = \begin{cases} k_{cone}|\tanh\left(\tan\left(\frac{\pi}{2}\frac{\theta_{sL}-\theta_{rt}}{\theta_{sL}-\theta_{sl}}\right)\right)|, & \theta_{rt} \geq \theta_{sl}, \\ 0, & \theta_{rt} < \theta_{sl}, \end{cases} \tag{6.93}$$

where $k_{cone} > 0$ is the elasticity coefficient, θ_{sl}, θ_{sL}, θ_{sr} and θ_{rR} are the inner and outer safety boundaries of the line-of-sight cone on the left and right sides, θ_{lt} and θ_{rt} are the line-of-sight angles on the left and right sides of the target relative to the active docking unit modality coordinate system. When θ_{lt} or θ_{rt} starts to enter the range of $[\theta_{sl}, \theta_{sL}]$ or $[\theta_{sR}, \theta_{sr}]$, the repulsive force starts to act, forcing the line-of-sight angles on both sides of the active docking unit modality to re-enter the safe range of $[\theta_{sr}, \theta_{sl}]$.

By introducing the force to stimulate the adjustment of the reference attitude, referring to the force distribution method, the vertical and lateral repulsive forces are designed as

$$\begin{aligned} \Delta_{cx} &= \left(F_{cone}^{lR} + F_{cone}^{lL}\right)\cos\theta_{lt} + \left(F_{cone}^{rR} + F_{cone}^{rL}\right)\cos\theta_{rt}, \\ \Delta_{cy} &= \left(F_{cone}^{lR} + F_{cone}^{lL}\right)\sin\theta_{lt} + \left(F_{cone}^{rR} + F_{cone}^{rL}\right)\sin\theta_{rt}, \end{aligned} \tag{6.94}$$

where Δ_{cx} is the longitudinal force, and Δ_{cy} is the lateral force.

By generating adaptive compensation terms through the auxiliary dynamic system to correct the reference trajectory in real time, a new safe guidance trajectory is generated, specifically as follows

$$\begin{aligned} \dot{\lambda}_{1cx} &= -k_{1cx}\lambda_{1cx} + \lambda_{2cx}, \\ \dot{\lambda}_{2cx} &= -k_{2cx}\lambda_{2cx} + \Delta_{cx}, \\ \dot{\lambda}_{1cy} &= -k_{1cy}\lambda_{1cy} + \lambda_{2cy}, \\ \dot{\lambda}_{2cy} &= -k_{2cy}\lambda_{2cy} + \Delta_{cy}, \end{aligned} \tag{6.95}$$

where k_{1cx}, k_{2cx}, k_{1cy}, and k_{2cy} are positive adaptive gains, and λ_{1cx}, λ_{2cx}, λ_{1cy}, and λ_{2cy} are adaptive auxiliary signals; λ_{1cx} and λ_{1cy} are equivalent to position correction signals, their boundedness and convergence are similar to the above section, which can be derived from Lemma 6.2, so the adaptive compensation terms are designed as $\Delta_{x_d} = \lambda_{1cx}$ and $\Delta_{y_d} = \lambda_{1cy}$.

6.7.2 Trajectory Controller with Composite Nonlinear Feedback

The tracking controller designed in the near-approach stage to quickly track the guidance trajectory is a prerequisite for reliable docking. Inspired by the idea of Zhao et al. [12] to improve the system's convergence speed during the transient process with a nonlinear composite control method, on the basis of the far-distance linear control framework, a nonlinear control term with fast transient convergence is combined to achieve rapid convergence of the straight-line guidance trajectory, ensuring that the latch smoothly enters the positioning hole within a short distance.

Based on the velocity and acceleration nested saturation model (6.4) and its velocity distribution, the following adaptive auxiliary dynamic system with composite nonlinear term convergence is designed

$$\begin{aligned}\dot{\lambda}_{ix}^{k} &= \cosh^{2}\left(\lambda_{ix}^{k}\right)\left(-k_{ix}\lambda_{ix}^{k} - \frac{1}{\alpha_{\lambda}T_{\lambda}}e^{|\lambda_{ix}^{k}|^{-\alpha_{\lambda}}}\operatorname{sig}\left(\lambda_{ix}^{k}\right)^{1-\alpha_{\lambda}} + \Delta_{ix}^{k}\right)/\gamma_{ix},\\ \dot{\lambda}_{iy}^{k} &= \cosh^{2}\left(\lambda_{iy}^{k}\right)\left(-k_{iy}\lambda_{iy}^{k} - \frac{1}{\alpha_{\lambda}T_{\lambda}}e^{|\lambda_{ix}^{k}|^{-\alpha_{\lambda}}}\operatorname{sig}\left(\lambda_{iy}^{k}\right)^{1-\alpha_{\lambda}} + \Delta_{iy}^{k}\right)/\gamma_{iy},\end{aligned} \tag{6.96}$$

and

$$\begin{aligned}\Delta_{ix}^{k} &= (v_{is}^{k} - v_{i}^{k})\cos(\theta_{i}^{k}) - l_{i}^{k}(w_{is}^{k} - w_{i}^{k})\sin(\theta_{i}^{k}),\\ \Delta_{iy}^{k} &= (v_{is}^{k} - v_{i}^{k})\sin(\theta_{i}^{k}) + l_{i}(w_{is}^{k} - w_{i}^{k})\cos(\theta_{i}^{k}),\end{aligned} \tag{6.97}$$

where λ_{ix}^{k} and λ_{iy}^{k} are auxiliary dynamic variables, Δ_{x}^{k} and Δ_{y}^{k} are the saturation deviations of the control input in the close-range stage, $k_{ix}, k_{iy} \in \mathbf{R}_{+}$ are linear control gains, T_{λ} and $0 < \alpha_{\lambda} < 1$ are nonlinear control gains, $\gamma_{ix} > 0$ and $\gamma_{iy} > 0$ are regulation gains, and $\operatorname{sig}(\cdot) = |\cdot|^{\alpha_{\lambda}}\operatorname{sgn}(\cdot)$. As known from the transient and steady-state convergence strategies by Huang et al. [13], the nonlinear term $\frac{1}{T_{\lambda}}e^{|\cdot|^{-\alpha_{\lambda}}}\operatorname{sig}(\cdot)^{1-\alpha_{\lambda}}$ plays a dominant role in the rapid convergence of timing, and the combination of linear terms can ensure the accuracy of steady-state convergence while accelerating the velocity of transient convergence.

The tracking error in the close-range stage is set as

$$\begin{aligned}e_{ix}^{k} &= x_{i}^{k} - \bar{x}_{d}^{k} - \gamma_{ix}\tanh\left(\lambda_{x}^{k}\right),\\ e_{iy}^{k} &= y_{i}^{k} - \bar{y}_{d}^{k} - \gamma_{iy}\tanh\left(\lambda_{y}^{k}\right),\end{aligned} \tag{6.98}$$

A composite nonlinear segmented controller is designed to achieve rapid error convergence. When the error $|e_{ix}^{k}| \geq \varepsilon_{x}$ or $|e_{iy}^{k}| \geq \varepsilon_{y}$, the longitudinal and lateral controllers in the close-range stage are

$$
\begin{aligned}
u_{ix}^k &= -c_3 e_{ix}^k - \frac{1}{\alpha_x T_x} e^{|e_{ix}^k|^{-\alpha_x}} \operatorname{sig}\left(e_{ix}^k\right)^{1-\alpha_x} + \dot{\bar{x}}_d^k - k_{ix}\lambda_{ix}^k - \frac{1}{\alpha_\lambda T_\lambda} e^{|\lambda_{ix}^k|^{-\alpha_\lambda}} \operatorname{sig}\left(\lambda_{ix}^k\right)^{1-\alpha_\lambda}, \\
u_{iy}^k &= -c_3 e_{iy}^k - \frac{1}{\alpha_y T_y} e^{|e_{iy}^k|^{-\alpha_y}} \operatorname{sig}\left(e_{iy}^k\right)^{1-\alpha_y} + \dot{\bar{y}}_d^k - k_{iy}\lambda_{iy}^k - \frac{1}{\alpha_\lambda T_\lambda} e^{|\lambda_{iy}^k|^{-\alpha_\lambda}} \operatorname{sig}\left(\lambda_{iy}^k\right)^{1-\alpha_\lambda},
\end{aligned}
\tag{6.99}
$$

When the error $|e_{ix}^k| < \varepsilon_x$ or $|e_{iy}^k| < \varepsilon_y$, the longitudinal and lateral controllers in the close-range stage are

$$
\begin{aligned}
u_{ix}^k &= -c_3 e_{ix}^k + \dot{\bar{x}}_d^k - k_{ix}\lambda_{ix}^k - \frac{1}{\alpha_\lambda T_\lambda} e^{|\lambda_{ix}^k|^{-\alpha_\lambda}} \operatorname{sig}\left(\lambda_{ix}^k\right)^{1-\alpha_\lambda}, \\
u_{iy}^k &= -c_3 e_{iy}^k + \dot{\bar{y}}_d^k - k_{iy}\lambda_{iy}^k - \frac{1}{\alpha_\lambda T_\lambda} e^{|\lambda_{iy}^k|^{-\alpha_\lambda}} \operatorname{sig}\left(\lambda_{iy}^k\right)^{1-\alpha_\lambda},
\end{aligned}
\tag{6.100}
$$

where $c_3 > 0$ is the controller gain.

The stability is analyzed in two cases. When $|e_{ix}^k| \geq \varepsilon_x$ or $|e_{iy}^k| \geq \varepsilon_y$, the Lyapunov functions $V_{3ix} = |e_{ix}^k|$ and $V_{3iy} = |e_{ix}^k|$ are chosen, and their derivatives are easily obtained as

$$
\begin{aligned}
\dot{V}_{3ix} &= \operatorname{sgn}\left(e_{ix}^k\right)\left(-c_3 e_{ix}^k - \frac{1}{\alpha_x T_x} e^{|e_{ix}^k|^{-\alpha_x}} \operatorname{sig}\left(e_{ix}^k\right)^{1-\alpha_x}\right) \leq -\frac{1}{a_x T_x} e^{|e_{ix}^k|^{-\alpha_x}} |e_{ix}^k|^{1-\alpha_x}, \\
\dot{V}_{3iy} &= \operatorname{sgn}\left(e_{iy}^k\right)\left(-c_3 e_{iy}^k - \frac{1}{\alpha_y T_y} e^{|e_{iy}^k|^{-\alpha_y}} \operatorname{sig}\left(e_{iy}^k\right)^{1-\alpha_y}\right) \leq -\frac{1}{a_y T_y} e^{|e_{iy}^k|^{-\alpha_y}} |e_{iy}^k|^{1-\alpha_y},
\end{aligned}
\tag{6.101}
$$

where $k_{ix} > 0$ and $k_{iy} > 0$ are linear control gains, and $0 < \alpha_x < 1$, $0 < \alpha_y < 1$, $T_x > 0$ and $T_y > 0$ are nonlinear gains related to rapid convergence.

When $|e_{ix}^k| < \varepsilon_x$ or $|e_{iy}^k| < \varepsilon_y$, we choose the Lyapunov functions $V_{3ix} = \frac{1}{2}\left(e_{ix}^k\right)^2$ and $V_{3iy} = \frac{1}{2}\left(e_{iy}^k\right)^2$. Their derivatives are easily obtained as

$$
\dot{V}_{3ix} = -c_3 e_{ix}^2, \ \dot{V}_{3iy} = -c_3 e_{iy}^2. \tag{6.102}
$$

From Eq. (6.101), we know that the controller (6.99) can make the error rapidly converge to the region where $|e_{ix}^k| < \varepsilon_x$ and $|e_{iy}^k| < \varepsilon_y$ within a predetermined time. Combined with Eq. (6.102), without considering the velocity error, the subsequent controller (6.100) will make the error continue to converge to 0.

6.8 Simulation and Experimental Verification

6.8.1 Dynamic Control Analysis of Simulation and Experiment

In this section, the lower layer dynamics of the reconfigurable swarm unit is taken as the object, and the robustness of the lower layer control and the advantage of MESO fusion estimation peak suppression are verified through simulation and experiment.

(1) Analysis of the Results of the Lower Layer Dynamics Simulation

The reference wheel velocity of the unit modality is set to $\omega_{dir}(t) = 300\sin(0.2\pi t)$ rad/s. The number of observers fused by MESO is set to 3. The wheel velocity state, the initial states of LESO and MESO fusion are respectively $\omega_{ir} = 0$ rad/s, $\hat{\omega}_{ir} = 100$ rad/s, $\hat{\omega}_{i1r} = 80$ rad/s, $\hat{\omega}_{i2r} = -60$ rad/s and $\hat{\omega}_{i3r} = 100$ rad/s. The initial values of uncertainty estimation for LESO and MESO adaptive fusion are respectively set to $\hat{d}_{ir} = 300$ N · m, $\hat{d}_{i1r} = 70$ N · m, $\hat{d}_{i2r} = -60$ N · m and $\hat{d}_{i3r} = 300$ N · m. The initial values for low-frequency adaptive learning are $\hat{W} = [0, 0]^{\mathrm{T}}$ and $\hat{W}_f = [0, 0]^{\mathrm{T}}$. The gain parameters for single ESO and LESO in MESO adaptive fusion are set to $\epsilon = 0.005$, $\kappa_1 = 2$ and $\kappa_2 = 1$. The design parameters for adaptive learning are $\sigma = 0.5$, $\gamma = 20$ and $\gamma_f = 0.5$, and the controller gain $k_{i\omega}$ is set to 0.5.

The comparison results of the wheel velocity tracking controller for the reconstructable swarm unit modality are shown in Fig. 6.6. The disturbance feedforward compensation structure generates excessive torque in the initial stage, which may directly affect the transient performance of the wheel velocity tracking of the unit modality. In Fig. 6.6a, the controller that combines MESO adaptive fusion estimation has smoother initial transient tracking behavior of the wheel velocity system compared to the controller that includes LESO (the initial peak values are 20 and 58 respectively). As can be seen from Fig. 6.6b, c, compared with LESO, the undesired peaks that appear in the transient process of MESO adaptive fusion can be ignored (the disturbance estimation peak values are −180 and −27 respectively). This shows that the MESO adaptive fusion estimation method proposed in this book can effectively suppress the peak phenomenon caused by the initial state error, so it can take larger gains in practical applications to achieve better robustness of the underlying dynamics of the unit modality. In Fig. 6.6d, the controller that integrates MESO also generates smoother control torque in the feedforward structure.

(2) Analysis of the Experimental Results of the Bottom Dynamics

The experimental setup for the active docking unit of the reconfigurable swarm uses the dynamic model shown in Eq. (6.65). To highlight the effect of disturbance estimation, the nonlinear model other than viscous friction under no-load wheel velocity is taken as the unknown uncertainty. The comparison of the fitting results of the unknown disturbance through the fusion estimation of MESO and the extended state of LESO verifies the advantages of the proposed method in disturbance and

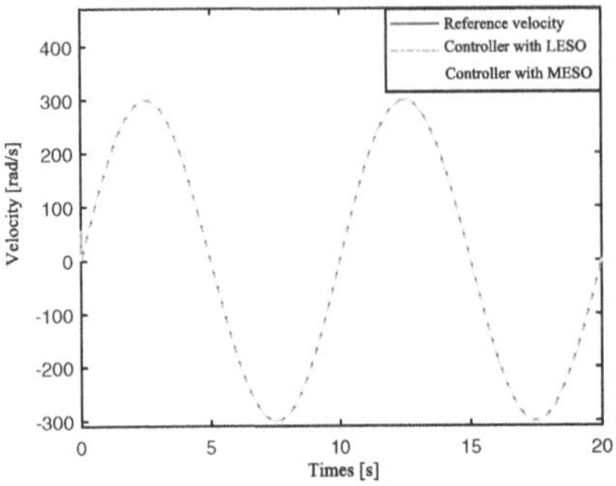

(a) Unit wheel velocity tracking comparison

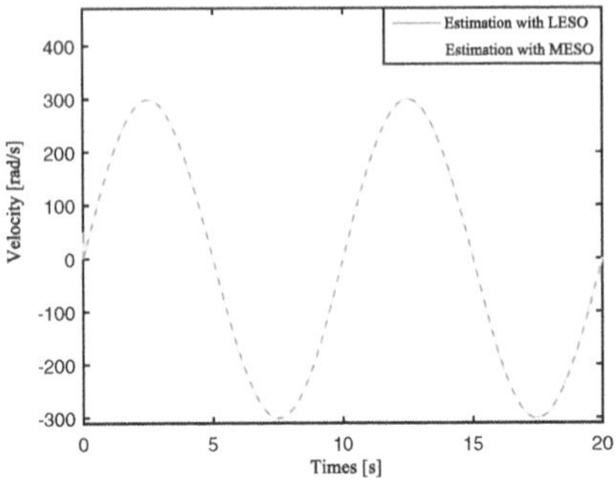

(b) Unit modality velocity estimation comparison

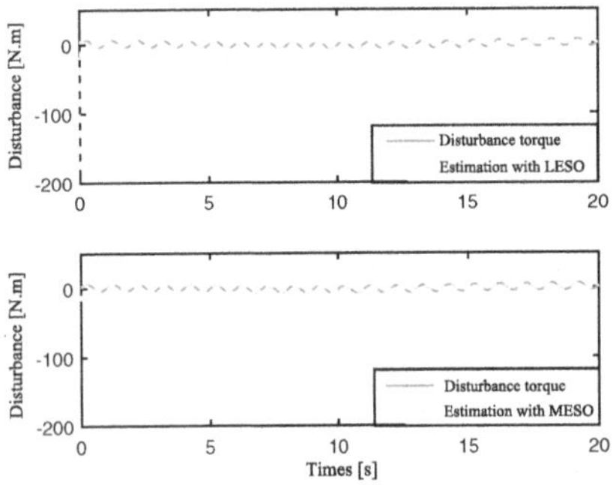

(c) Unit disturbance estimation comparison

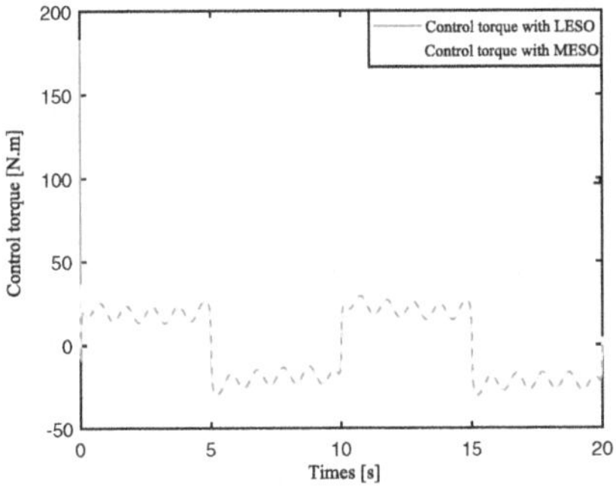

(d) Unit control torque comparison

Fig. 6.6 Simulation comparison of unit modality dynamic controller effect of MESO and LESO

estimation peak suppression. The experimental reference wheel velocity of the right wheel of the active docking unit is

$$\omega_{dir} = \sum_{l=1}^{N} a_l \cos(\omega l t) + b_l \sin(\omega l t) \text{ rad/s},$$

where $\omega = \frac{\pi}{5}$, and the corresponding amplitudes are $a_1 = -0.9378, b_1 = 4.5896, a_2 = -0.1495, b_2 = -0.3939, a_3 = 0.1683, b_3 = 0.1479, a_4 = -0.1271, b_4 = 0.0051, a_5 = 0.0443, b_5 = -0.0513$. The total disturbance $f_d = f_c(\omega_{ir}) + d_{ir}$ N.m represents unknown uncertainties and other disturbances; the estimated value of the total disturbance $\hat{f}$ can be obtained through ESO or MESO fusion. In the experiment, two schemes for estimating the total disturbance were compared, namely, the comparison verification of LESO and MESO adaptive fusion estimation of the total disturbance. These two comparative experiments used the same controller gains and observer gains, and the initial state settings were the same.

In Fig. 6.7, the comparison of MESO adaptive fusion and LESO aggregated disturbance estimation and reference wheel velocity tracking are shown. In Fig. 6.7a,

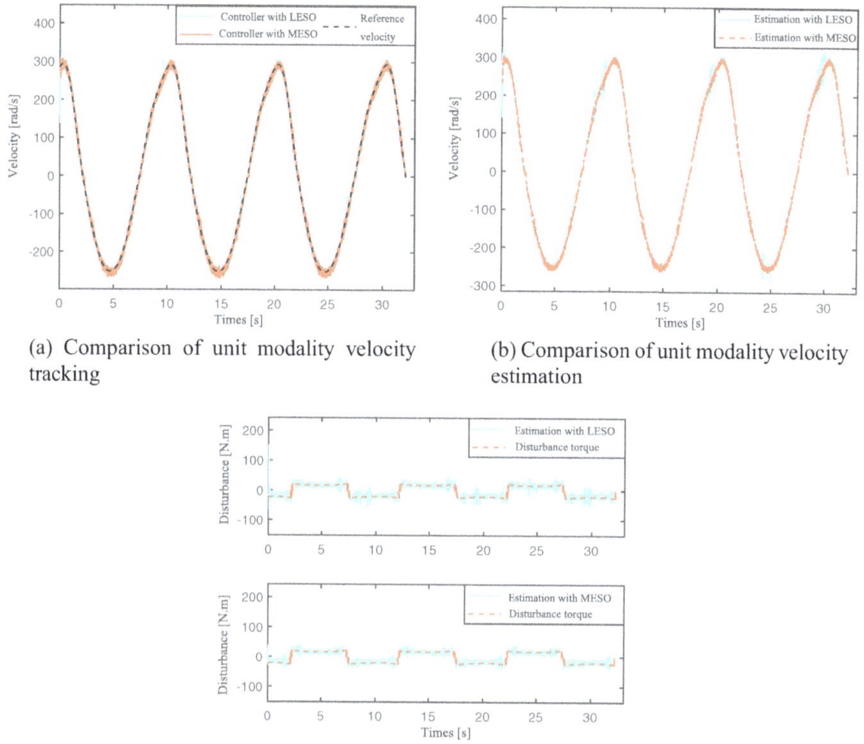

(a) Comparison of unit modality velocity tracking

(b) Comparison of unit modality velocity estimation

(c) Comparison of unit modality disturbance estimation

Fig. 6.7 Experimental comparison of unit modality dynamic controller effect of MESO and LESO

due to the feedforward compensation of the aggregated disturbance, the transient overshoot values of the fusion MESO and LESO control schemes are 28 and 148.8 respectively, showing obvious smoothness advantages. MESO has shown significant improvements in the controller synthesis for balance estimation accuracy and reducing peak phenomena. In terms of peak suppression of aggregated disturbance estimation, the MESO fusion in Fig. 6.7b, c shows significant advantages over the traditional LESO, where the peak values of disturbance estimation for the two observers are 149.897 (LESO) and 28.9526 (MESO) respectively. This improvement makes it possible to more flexibly implement gain scheduling in disturbance estimation. Therefore, this method shows significant feasibility in reducing tracking errors and optimizing the active docking of unit modality or the entire reconfigurable swarm velocity tracking in transient and steady-state stages of response dynamics.

6.8.2 *Kinematic Docking Analysis of Simulation and Experiment*

(1) Analysis of Simulation Results of Unit Mode Stage-by-Stage Combination Docking

This section aims to verify the effectiveness of the hierarchical control strategy in kinematics. An active docking unit kinematics and dynamics model is built on the MATLAB R2021b simulation platform. Based on this, multiple sets of hierarchical control strategies are deployed for comparison, verifying the application advantages of the proposed hierarchical control strategy in long-distance safe assembly and close-range precise docking. The initial position of the active docking unit is $x_i(0) = -0.2\,\text{m}$, $y_i(0) = -0.2\,\text{m}$ and $\theta_i(0) = 0\,\text{rad}$, and the position of the docking target unit is $x_k = 6.1\,\text{m}$, $y_k = 7.2\,\text{m}$ and $\theta_k = \frac{\pi}{2}\,\text{rad}$. Therefore, a circular long-distance assembly reference trajectory and a straight-line guidance trajectory located behind the docking target are set in the simulation, guiding the active docking unit to complete the large-scale unit combination docking. The saturation constraints and control gains of the unit modality are similar to those of the swarm control.

In response to the multi-dimensional constraints that may occur during the far/near stage process, the controller is designed to achieve adaptive long-distance assembly and close-range docking, and in the experimental design aspect, (1) the global positioning information is used to construct the inertial coordinate system and the stationary coordinate system with the docking target as the origin; (2) the line-of-sight angle and size characteristics of the active unit modality under the docking target system are simulated to constrain the line of sight; (3) the nested motion saturation constraint is introduced to limit the motion performance of the unit modality; (4) environmental obstacles are introduced to threaten the safe movement of the unit modality, so as to fully reflect the characteristics of real application scenarios and support the physical verification of the proposed hierarchical reconfiguration algorithm in a limited experimental field and equipment.

The existing unit modality combination docking algorithms mainly focus on the implementation of reconfiguration functions, mostly open-loop and simple closed-loop strategies, and it is difficult to successfully deploy the docking function under multiple constraints set in the sub-stages. The docking of the unit modality is essentially a safe control problem under multi-dimensional constraints. Therefore, the trajectory control and safety strategy proposed by Wang et al. [11] are adopted in the active docking unit modality near and far distance stages. They are marked as C1 and C2 respectively, and the proposed algorithm is compared and verified for its smoothness and safety advantages in the experimental setting scenario. The near-distance guidance stage uses the proposed control method, the composite nonlinear control method that removes the line-of-sight safety constraint, and the single linear feedback control method, which are marked as C1, C2, and C3 respectively, to compare and verify the advantages of the proposed near-distance control method in precise docking and line-of-sight safety. In addition, the method proposed in this chapter is used at all levels of the unit modality dynamics.

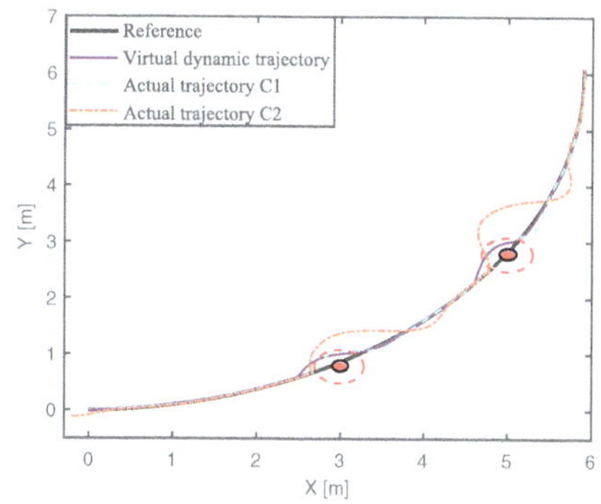

(a) Comparison of planar trajectories in the far-distance stage

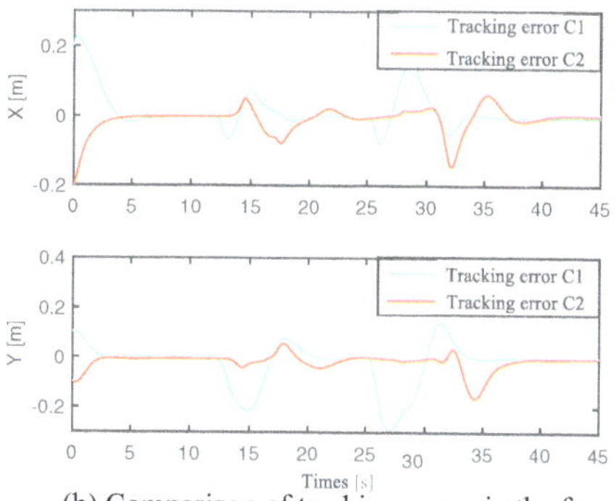

(b) Comparison of tracking errors in the far-distance stage

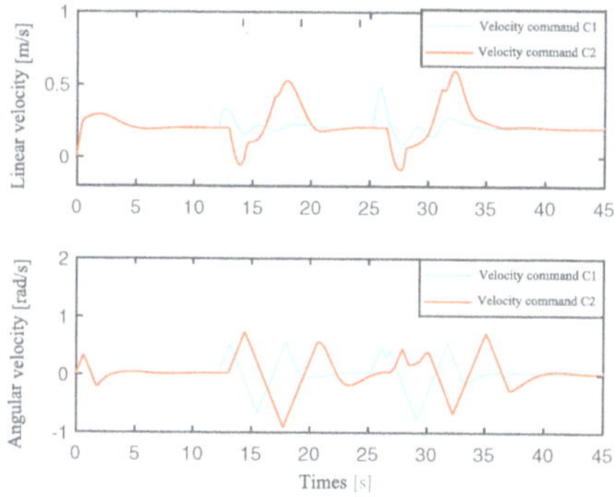

(c) Long-range velocity command comparison

Fig. 6.8 Simulation results in the long-range assembly phase of unit modality

① Simulation verification of the far-distance stage: Fig. 6.8 shows the comparison of trajectory control effects in the far-distance assembly stage. Although in 6.8b, the two comparison methods have quite accurate control precision in the steady state, the upper-level safety barrier and the lower-level virtual spring safety strategy proposed in this book are shorter in avoiding obstacles in Fig. 6.8a and have smaller amplitudes, as well as more gentle velocity control commands in Fig. 6.8c compared to the single force safety mechanism proposed by Wang. Therefore, the proposed far-distance assembly control method is more suitable for complex engineering application scenarios.

② Close-range simulation experiment: Fig. 6.9 shows the comparison results of the trajectory tracking in the close-range phase. As can be seen from Fig. 6.9a–b, the control scheme with the composite nonlinear exponential convergence term can achieve a faster convergence velocity than the linear control scheme, ensuring that the active docking unit can smoothly deliver the lock into the positioning hole within a short distance, completing the locking reconfiguration of different units. At the same time, as can be seen from Fig. 6.9d, the composite nonlinear feedback mode

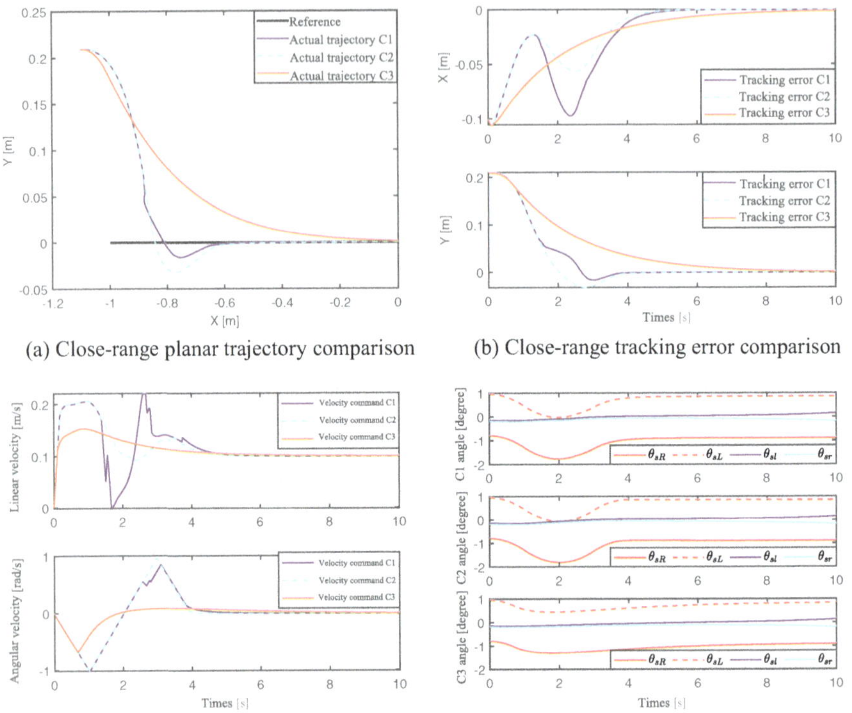

(a) Close-range planar trajectory comparison (b) Close-range tracking error comparison

(c) Close-range velocity command comparison (d) Close-range line-of-sight safety comparison

Fig. 6.9 Simulation results in the close-range docking phase of unit modality

will cause greater changes in the posture of the active docking unit, causing the active docking unit to lose sight of the recognition target, and the introduction of the line-of-sight correction mechanism of the reference trajectory can drive the active docking unit to always recognize the target during the docking process, ensuring the line-of-sight safety of the entire close-range journey.

(2) Analysis of the Experimental Results of the Unit Modality Docking in Stages
The experimental platform built in Chap. 2 was used to conduct the modular unit combination docking experiments. The pose of each unit modality is obtained by collecting and recognizing the Tag through the overhead camera, and a local area communication network between the host computer and each unit modality is established based on the WiFi module and router, realizing information sharing among each unit modality, real-time status monitoring and motion control of the host computer ROS. The unit modality is subject to nested saturation constraints on velocity and acceleration, with saturation thresholds set to $\bar{v} = 1.5\ \mathrm{m/s}$, $\bar{a} = 1\ \mathrm{m/s^2}$, $\bar{w} = 1.5\ \mathrm{rad/s}$ and $\bar{\omega} = 1\ \mathrm{rad/s^2}$. The maximum control error limit for successful reconfiguration of the docking device lock and positioning hole is 2 cm.

① Remote stage assembly experiment verification

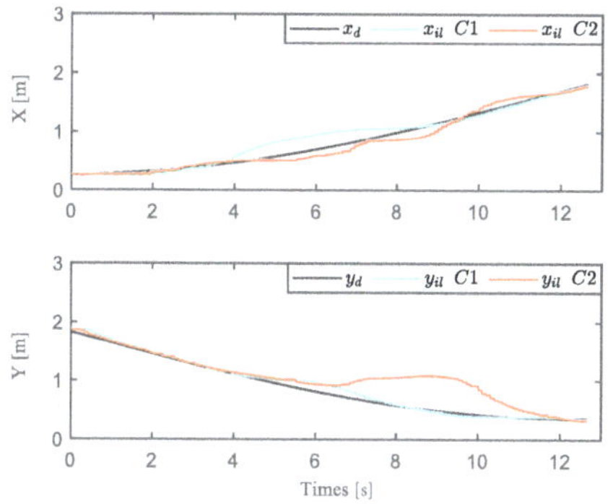

(a) Long-range trajectory tracking performance

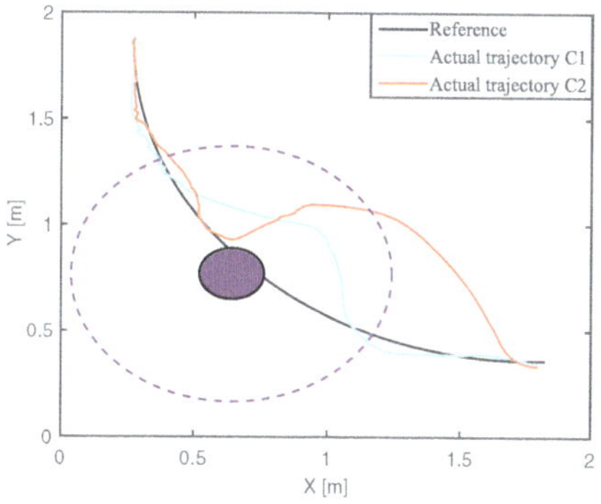

(b) Long-range planar aggregation trajectory

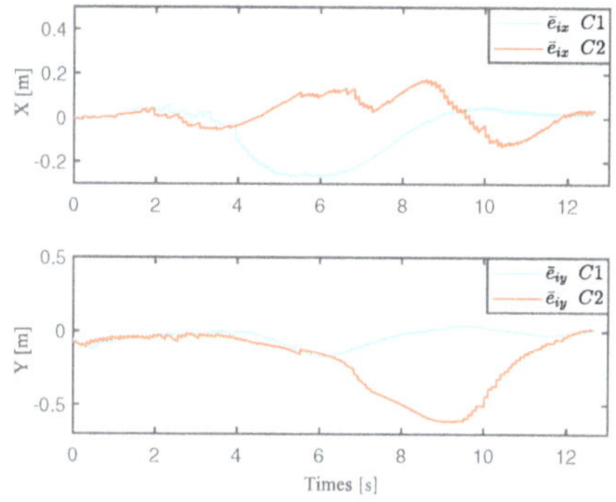

(c) Long-range trajectory tracking error

Fig. 6.10 Experimental results in the long-range assembly phase of unit modality

In the remote stage, a circular trajectory is designed based on the global pose of the unit modality and the target to be docked, so that the unit modality moves to the position directly behind the unit modality to be docked. At the same time, an obstacle $p_o = [x_o, y_o]^{\mathrm{T}}$ is set on the desired trajectory as shown in Fig. 6.10 to verify the safety and smoothness of the proposed assembly trajectory tracking method. The position of the obstacle can be obtained by recognizing the Tag attached to the surface of the obstacle with the camera. To ensure the fairness of the comparison between the two methods in the remote stage experiment, the expected trajectories, initial positions of the modular units, collision avoidance distances, and control gains are set to be the same for both experiments. The controller gains are designed as $c_1 = c_2 = c_3 = 1$, the adaptive gains are designed as $k_{ix} = k_x^{\varphi} = k_{iy} = k_y^{\varphi} = 10$ and $k_{sx} = k_{sy} = 30$. The safety barrier gains are designed as $\alpha = 20$, $\gamma = 0.4$ and $\rho = 0.5$, the collision avoidance distance is set to 0.6 m, the inner and outer boundaries of the virtual spring are $\rho_{sr} = 0.15$ m and $\rho_{sR} = 0.15$ m, the gain $k_r = 0.5$, and the safety constraint boundaries before and after obstacle avoidance are $x_o^u = y_o^u = 3$ m,

$x_o^l = y_o^l = 0$ m and $x_o^u = y_o^u = 3$ m, $y_o^l = 0$ m, $x_o^l = 0.84$ m. To ensure the continuity of the boundaries, the switching boundaries can be filtered.

This chapter is based on the safety barrier generation integrated with the nested saturation of the upper layer obstacle avoidance guidance trajectory, which plays a leading role in the process of avoiding obstacles, and ensures complete safety within the range near the obstacle through the lower layer's virtual spring mechanism. The hierarchical safety constraint mechanism has a clear stability advantage over the APF reactive obstacle avoidance method proposed by Sharma et al. [11] in situations where control input saturation is easily triggered. Figure 6.10a, c reflect the transient response and steady-state accuracy of the two methods in the trajectory assembly, in terms of tracking accuracy, the proposed method (C1: 0.029 m) is basically consistent with the comparison method (C2: 0.031 m), but its oscillation amplitude and recovery time for avoiding obstacles are 0.3 m and 3 s, respectively, which have better smooth characteristics compared to the oscillation amplitude and recovery time of C2 (0.5 m and 6 s). Figure 6.10b is the planar trajectory of the long-distance assembly process, the actual tracking trajectory of C1 is shorter than the obstacle avoidance path of C2, and the response is smoother, thus it is more suitable for complex reconfiguration scenes with multiple obstacle distributions.

② Verification of the close-range assembly experiment In the near interception phase, a straight-line guidance trajectory is designed based on the target to be intercepted, and the active interceptor unit is driven by local sensing feedback to accurately track the guidance trajectory with its dual wheels and move to the target position. This section constructs a simulation of the posture and line of sight of local sensing feedback based on the global posture and coordinate transformation of the active interceptor unit and the target to be intercepted, and designs a guidance trajectory tracking controller based on the proposed method and the trajectory tracking method proposed by Sharma et al. [11], which also introduces a line of sight safety constraint in the comparative controller but does not introduce adaptive anti-saturation.

To ensure the fairness of the comparison, the control gains of the two experiments are both set to $c_4 = 0.5$. The proposed method also includes nonlinear gains $\alpha_x = \alpha_y = 0.4$ and $T_x = T_y = 4$, and adaptive gains $k_{ix} = k_{iy} = 10$, $\alpha_\lambda = 0.4$ and $T_\lambda = 4$. The design parameters for the line of sight safety constraint are $k_{cone} = 0.5$, $k_{1cx} = k_{1cy} = k_{2cx} = k_{2cy} = 30$. Figure 6.10a shows the trajectory tracking effect of the combined docking process. In the initial stage, the position error can cause a sudden change in posture, triggering the line-of-sight safety mechanism to correct the reference trajectory to ensure that the docking target is always within the safe line of sight. The control strategy of combined nonlinear feedback promotes the docking trajectory to converge accurately and quickly to the guidance trajectory to ensure that the positioning hole is fed into the lock. Figure 6.10b also shows the smoothness and safety of C1 during the docking process. However, C2, because it does not introduce adaptive anti-saturation, is prone to trigger the saturation threshold when triggering the line-of-sight safety mechanism. The movement trajectory of the mobile unit in Fig. 6.10a, b is uncontrolled and oscillates severely, causing frequent corrections to the guidance trajectory and ultimately failing to complete the close-range docking smoothly. Figure 6.10c shows the tracking error of the docking guidance trajectory

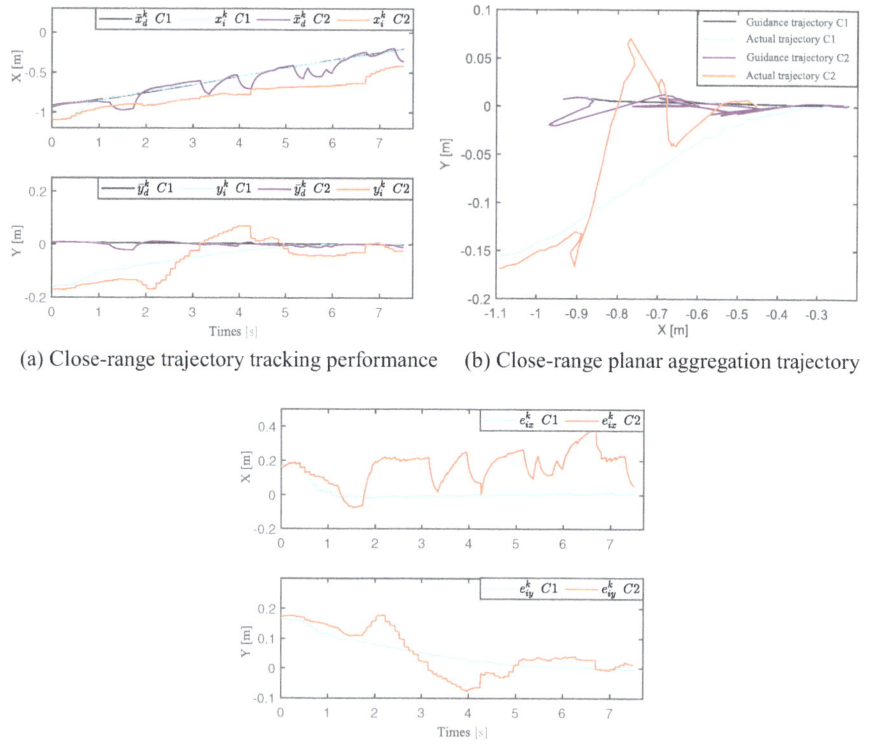

(a) Close-range trajectory tracking performance (b) Close-range planar aggregation trajectory

(c) Close-range trajectory tracking error

Fig. 6.11 Experimental results in the close-range docking phase of unit modality

in the close-range stage. With a docking accuracy requirement of 0.03 m, the time for C1 to reach the range of docking accuracy is 4 s, and the final control accuracy at the positioning hole is 0.007 m, while C2 is in a state of loss of control (Fig. 6.11).

6.9 Summary

This book addresses the demand for reconfiguration in a broad task environment, proposing a segmented secure docking control framework based on long-distance global positioning and short-distance local sensing. In the long-distance range, a hierarchical security constraint mechanism and self-adaptive anti-saturation are introduced to ensure the smoothness and safety of the assembly. In the short-distance range, a line-of-sight-based self-adaptive trajectory compensation term and a composite nonlinear feedback controller are designed to achieve smooth and rapid docking within the line-of-sight range. At the same time, a MESO disturbance fusion estimation that suppresses transient peaks is introduced into the bottom dynamics to enhance the disturbance resistance of velocity control. Finally, through comparative simulations and experiments of combination docking, the effectiveness of the dynamic disturbance control and subsequent docking algorithm stages is verified, ensuring reliable switching from the unit modality to the combined modality.

References

1. Huang, B., Zhou, B., Pang, H., Xu, J., Fang, H.: A reliable docking mechanism and close-range docking algorithm for modular reconfigurable robots. In: Proc. 2023 9th Int. Conf. Mechatronics Robot. Eng. (ICMRE), pp. 78–83 (2023)
2. Tan, W., Wei, H., Yang, B.: SambotII: A new self-assembly modular robot platform based on Sambot. Appl. Sci. **8**(10), 1719 (2018)
3. Li, H., Wang, H., Cui, L., Li, J., Wei, Q., Xia, J.: Design and experiments of a compact self-assembling mobile modular robot with joint actuation and onboard visual-based perception. Appl. Sci. **12**(6), 3050 (2022)
4. Quiñonez, Y., Baca, J., Delope, J., Ferre, M., Aracil, R.: Self-alignment approach based on cooperative behaviors for the docking process of modular mobile robots. In: Proc. 2010 IEEE Electron. Robot. Automot. Mech. Conf. (CERMA), pp. 445–450 (2010)
5. Wei, H.X., Li, H.Y., Guan, Y., Li, Y.D.: A dynamics based two-stage path model for the docking navigation of a self-assembly modular robot (Sambot). Robotica **34**(7), 1517–1528 (2016)
6. Ghaffari, A.: Analytical design and experimental verification of geofencing control for aerial applications. IEEE/ASME Trans. Mechatron. **26**(2), 1106–1117 (2020)
7. Ran, M., Li, J., Xie, L.: A new extended state observer for uncertain nonlinear systems. Automatica **131**, 109772 (2021)
8. Shakarami, M., Esfandiari, K., Suratgar, A.A., Talebi, H.A.: Peaking attenuation of high-gain observers using adaptive techniques: State estimation and feedback control. IEEE Trans. Autom. Control **65**(10), 4215–4229 (2020)
9. Yucelen, T., Haddad, W.M.: Low-frequency learning and fast adaptation in model reference adaptive control. IEEE Trans. Autom. Control **58**(4), 1080–1085 (2013)
10. Guo, B., Zhao, Z.: On the convergence of an extended state observer for nonlinear systems with uncertainty. Syst. Control Lett. **60**(6), 420–430 (2011)
11. Sharma, R.S., Mondal, A., Behera, L.: Tracking control of mobile robots in formation in the presence of disturbances. IEEE Trans. Ind. Inform. **17**(1), 110–123 (2021)
12. Zhao, K., Zhang, J., Ma, D., Xia, Y.: Composite disturbance rejection attitude control for quadrotor with unknown disturbance. IEEE Trans. Ind. Electron. **67**(8), 6894–6903 (2019)
13. Huang, S., Xiong, L., Zhou, Y., Liu, J., Jia, Q., Li, P.: A novel distributed predefined-time sliding mode controller for performance enhancement of power system under input saturation. IEEE Trans. Circuits Syst. I, Reg. Pap. **69**(10), 4284–4297 (2022)
14. Wang, W., Yu, W., Zhang, H.: JL-2: A mobile multi-robot system with docking and manipulating capabilities. Int. J. Adv. Robot. Syst. **7**(1), 9–18 (2010)
15. Wang, W., Lei, Z.: Trajectory planning for a reconfigurable mobile robot in mobile manipulating mode. In: Proc. 11th World Congr. Intell. Control Autom. (WCICA), pp. 42–47 (2014)
16. Zhao, D., Luo, H., Tu, Y., Meng, C., Lam, T.: Snail-inspired robotic swarms: A hybrid connector drives collective adaptation in unstructured outdoor environments. Nat. Commun. **15**, 3647 (2024)

Chapter 7
Modeling and Anti-disturbance and Fault-Tolerance Control of Combined Modalties

Abstract The adaptive enhancement control framework for reconfigurable swarm in combined modality model under actuator failures, unknown disturbances, and safety saturation constraints is designed in this chapter. A kinematic and dynamic model adaptive with failures and uncertain disturbances is constructed. A kinematic control method combining high-level composite nonlinear feedback with adaptive safety boundaries is proposed, while the bottom dynamic controller integrates the noise-smooth and low-power disturbance estimation as well as adaptive fault-tolerant mechanism, achieving robust velocity tracking under faults and disturbances. Simulations and experiments verify the fault-tolerance, disturbance rejection, and safety of the proposed approach.

Keywords Combination modality · Safety boundary · Noise and peaking rejection · Fault tolerance

7.1 Introduction

The combined modality of a mobile connected structure after multiple modular unit reconfiguration is a key feature distinguishing it from traditional mobile vehicle swarms. However, there is a general lack of research results related to model construction and internal control of the modal. Lashkari et al. [1] constructed a kinematic and dynamic model of a vehicle connected by two units with a rear unit failure, and realized the control of the dual-unit combined modality with a rear unit failure using a kinematic and dynamic level control structure. However, the model essentially uses a tractor-trailer model, which is not applicable to combined modality models with more than 2 healthy units. Shi et al. [2] used a hierarchical control framework to achieve instantaneous rotation center positioning and steering feedback, but lacked precise control of the bottom dynamics. Leggieri et al. [3] used DH parameters and Lagrange equations to construct kinematic and dynamic models of any combination scale, but lacked matching and efficient control strategies. Modeling and control of other types of reconfigurable objects, such as connected unmanned surface vehicle formations [4] and quadrotor swarms [5], have not considered the negative impact of

T. Jiang et al., *Multi-modality Control of Reconfigurable Swarm*,
https://doi.org/10.1007/978-981-95-0305-6_7

application limitations in complex environments and uncertainties within and outside the combined modality model. Therefore, existing results generally lack practical research methods for modality control of connected modular units and strategies for enhancing adaptability in complex environments.

The reconfigurable swarm combination modality has the redundant actuation advantage over the single modular unit, not only can it recover the mobility ability through the reconfiguration of the healthy unit/combination members and the faulty unit/combination members, but it can also maintain the mobility performance under the random occurrence of combination modality member faults by applying the self-adaptive fault tolerance mechanism. Therefore, this chapter introduces the random actuation fault synchronously under the constraints of saturation and safety and the negative interference of lumped uncertainties, and designs a robust controller based on the kinematics-dynamics cascaded structure of the combined modality to achieve strong adaptability of the combined modality member trajectory control under actuation faults, unknown uncertainties, and environmental constraints. Firstly, at the kinematics level, a safety avoidance mechanism based on self-adaptive boundary compensation is proposed, and a combination modality member kinematics trajectory controller is designed based on composite nonlinear feedback. Secondly, at the dynamics level, by introducing the Adaptive Gain Filter Extended State Observer (AGFESO), it can smoothly recover the disturbance signal from the noisy output under the condition of large initial estimation error, and enhance the disturbance resistance performance of the bottom velocity control through the feedforward compensation structure; at the same time, the self-adaptive fault tolerance mechanism based on the optimization strategy is introduced to make the torque distribution of the control input match the actual actuation faults, avoiding the performance degradation and collapse of the combination modality when the fault randomly occurs. Finally, through multiple sets of simulations and experiments, the adaptive enhancement effect of the proposed method on fault tolerance, disturbance rejection, and safety performance is verified.

7.2 Combination Modality Modeling and Problem Description

7.2.1 Combination Modality Kinematic and Dynamic Model Construction

The multiple modular units designed in this book can be reconfigured into a chain-type combination modality of any number and scale by the docking mechanism. The main advantages of multi-unit assembly reconfiguration are self-recovery and enhanced passability. The self-recovery capability is mainly reflected in the combination of healthy members and faulty members. The reconfigurated combination modality member can restore its mobility, which is more task-resilient compared to traditional mobile vehicles. The enhancement of passability is mainly reflected in the autonomous reconfiguration of n healthy units/members (or m faulty units and $n-m$ healthy units), which enhances the cross-terrain capability of the mobile swarm. This type of combination modality model can adaptively adjust the scale of the reconfigurable modular unit according to the obstacle terrain and passability requirements, and has obvious multimodal characteristics during task execution. In addition, although there is currently a lack of practical models of kinematics and dynamics of combination modality composed of differential drive modular units, the modeling process of combination modality such as unmanned surface vehicles and airial vehicles [4–6] can still be referenced. General modeling methods and theoretical tools can be used to carry out derivation and research on the kinematics and dynamics of combination modality of any rescongfigurable scale.

(1) Kinematic Model Construction of the Combination Modality

To simplify the control process of the combination modality member, as shown in Fig. 7.1, the control point is set at the center of the combination modality member, with the control point position set as $r_f = \left[x_f, y_f\right]^{\mathrm{T}}$, and the control point direction as $\theta_f = \theta_1$. Therefore, the corresponding combination modality control point position vector is $p_f = \left[x_f, y_f, \theta_f\right]^{\mathrm{T}}$. According to Assumption 7.1, each modular unit has the same linear velocity and angular velocity, and the control point linear velocity, angular velocity, and corresponding velocity vector are $v_f = v_1$, $w_f = w_1$, and $\eta_f = \left[v_f, w_f\right]^{\mathrm{T}}$ respectively.

Assumption 7.1 *The coupling mechanisms of each modular unit in the wheel-driven combined modality member are tightly connected, and there is no slippage or relative displacement caused by it between adjacent modular members.*

The offset of the center position between adjacent modular units in the modularized design is set as d, and the combination scale is m module. Knowing the position of a certain unit, the positions of all modular units can be deduced through structural

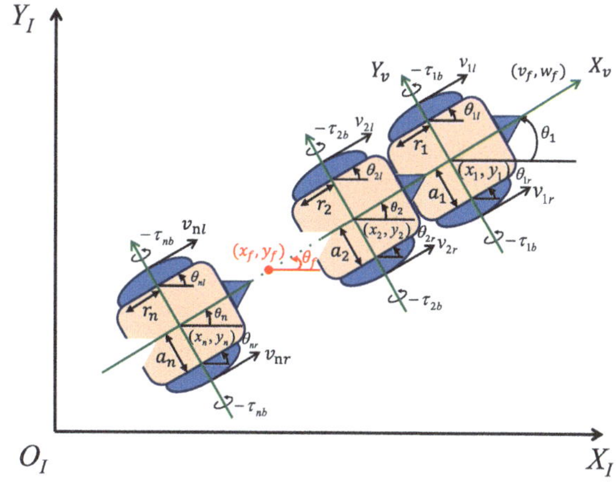

Fig. 7.1 Combination modality frame diagram

parameters and direction propagation. Therefore, according to Assumption 7.1, the position of the first unit is set as $p_{c1} = [x_1, y_1]^{\mathrm{T}}$, the position of the ith unit modality is $p_{ci} = \left[x_f - (i-1)\,d\cos\theta_f, y_f + (i-1)\,d\sin\theta_f\right]^{\mathrm{T}}, i = 1, \ldots, m$, then the direction of each unit in the combination modality member is $\theta_f = \theta_1 = \cdots = \theta_m$, and the position of the control point of the combination modality member is $p_f = \left[x_1 - \frac{1}{2}(m-1)\cos\theta_f, y_1 - \frac{1}{2}(m-1)\sin\theta_f\right]^{\mathrm{T}}$.

The control point of the combination modality member has the same nonholonomic constraints and kinematic model as the first unit, which is expressed as follows

$$\dot{p}_f = S\left(p_f\right)\eta_f, \tag{7.1}$$
$$A\left(p_f\right)\dot{p}_f = 0, \tag{7.2}$$

where,

$$S\left(p_f\right) = \begin{bmatrix} \cos\theta_f & 0 \\ \sin\theta_f & 0 \\ 0 & 1 \end{bmatrix}, \eta_i = \begin{bmatrix} v_f \\ w_f \end{bmatrix}, A\left(p_f\right) = \left[\sin\theta_f, -\cos\theta_f, 0\right].$$

(2) Construction of the Kinematic Model Construction for the Combination Modality

The kinematic model of the combination modality member will dynamically change with the dynamic reconfiguration process. It is necessary to construct a variable-dimensional kinematic equation that can be extended with the scale as the premise basis for kinematic control. Here, the Lagrange method is also used to derive the kinematic equation of the combination modality member after the reconfiguration of multiple units.

If the reconfiguration scale of the combination modality member is m unit, according to Assumption 7.1 and structural parameters, the position of the center of gravity of the first unit is $r_{c1} = \left[x_1 - a\cos\theta_f, y_f - a\sin\theta_f\right]^{\mathrm{T}}$, the position of the center of gravity corresponding to the ith unit is

$r_{ci} = \left[x_1 - a\cos\theta_f - (i-1)\,d\cos\theta_f, y_1 - a\sin\theta_f - (i-1)\sin\theta_f\right]^{\mathrm{T}}$, where each unit is a modular unit with basically the same shape and mass, so it is assumed that $a_i = a$. At this time, it is also easy to infer that the overall center of gravity of the combination modality member is $r_f = \left[x_1 - a\cos\theta_f - \frac{m-1}{2}d\cos\theta_f, y_1 - a\sin\theta_f - \frac{m-1}{2}d\sin\theta_f\right]^{\mathrm{T}}$. Their overall translational and rotational kinetic power can be written as

$$\begin{aligned} T_f &= \frac{1}{2}\sum_{i=1}^{m}\left(m_i\dot{r}_{ci}^{\mathrm{T}}\dot{r}_{ci} + I_{mi}\dot{\theta}_i^2\right) \\ &= \frac{1}{2}\sum_{i=1}^{m}\left(m_i\dot{p}_{ci}^{\mathrm{T}}T_i^{\mathrm{T}}T_i\dot{p}_{ci} + I_{mi}\dot{\theta}_i^2\right) \end{aligned}$$

$$
\begin{aligned}
&= \frac{1}{2}\dot{p}_f^{\mathrm{T}}\left(\operatorname{diag}\{0, 0, \sum_{i=1}^{m} I_{mi}\} + \sum_{i=1}^{m} m_i T_i^{\mathrm{T}} T_i\right)\dot{p}_f \\
&= \frac{1}{2}\dot{p}_f^{\mathrm{T}} M_f \dot{p}_f,
\end{aligned} \tag{7.3}
$$

where M_f is the inertia matrix of the combination modality member.

According to $T_i = \frac{\partial r_{ci}}{\partial p_f}$, the inertia matrix M_f of the combination modality dynamics can be obtained

$$
M_f = \begin{bmatrix} m_{f11} & m_{f12} & m_{f13} \\ m_{f21} & m_{f22} & m_{f23} \\ m_{f31} & m_{f32} & m_{f33} \end{bmatrix}, \tag{7.4}
$$

and

$$
\begin{aligned}
&m_{f11} = m_{f22} = \sum_{i=1}^{m} m_i, m_{f33} = \sum_{i=1}^{m} m_i a^2 + I_{mi}, m_{f12} = m_{f21} = 0, \\
&m_{f13} = m_{f31} = \sum_{i=1}^{m} m_i a \sin\theta_f, m_{f23} = m_{f32} = -\sum_{i=1}^{m} m_i a \cos\theta_f.
\end{aligned}
$$

From Eq. (7.4), the derivative of the inertia matrix M_f of the combination modality can be obtained as follows

$$
\dot{M}_f = \begin{bmatrix} \dot{m}_{f11} & \dot{m}_{f12} & \dot{m}_{f13} \\ \dot{m}_{f21} & \dot{m}_{f22} & \dot{m}_{f23} \\ \dot{m}_{f31} & \dot{m}_{f32} & \dot{m}_{f33} \end{bmatrix}, \tag{7.5}
$$

and

$$
\begin{aligned}
&\dot{m}_{f11} = \dot{m}_{f12} = \dot{m}_{f21} = \dot{m}_{f22} = \dot{m}_{f33} = 0, \\
&m_{f13} = m_{f31} = \sum_{i=1}^{m} m_i a \dot{\theta}_f \cos\theta_f, m_{f23} = m_{f32} = \sum_{i=1}^{m} m_i a \dot{\theta}_f \sin\theta_f.
\end{aligned}
$$

The centripetal force and Coriolis force terms of the kinematic model of the combination modality member are $E_f\left(p_f, \dot{p}_f\right) = \dot{M}_f \dot{p}_f - \frac{\partial T_f}{\partial p_f}$, where the elements of $\frac{\partial T_f}{\partial p_f}$ can be written as

$$
\begin{aligned}
&\frac{\partial T_f}{\partial x_f} = 0, \frac{\partial T_f}{\partial y_f} = 0, \\
&\frac{\partial T_f}{\partial \phi_f} = \sum_{i=1}^{m} m_i a \dot{\theta}_f \cos\theta_f \left(\dot{x}_f + a\sin\theta_f\right) + \sum_{i=1}^{m} m_i a \dot{\theta}_f \sin\theta_f \left(\dot{y}_f - a\cos\theta_f\right).
\end{aligned} \tag{7.6}
$$

From Eqs. (7.5) and (7.6), $E_f\left(p_f, \dot{p}_f\right)$ can be calculated as

$$E_f = \begin{bmatrix} \sum_{i=1}^m m_i a\dot{\theta}_f^2 \cos\theta_f \\ \sum_{i=1}^m m_i a\dot{\theta}_f^2 \sin\theta_f \\ 0 \end{bmatrix}. \tag{7.7}$$

The specific form of the generalized force vector of the combination modality $Q_f = \sum_{i=1}^m \left(\frac{\partial F_i}{\partial p_f} + \frac{\partial M_{ti}}{\partial p_f}\right)$ is as follows

$$Q_f = \begin{bmatrix} \sum_{i=1}^m \frac{1}{r}\left(\tau_{ir} + \tau_{il}\right)\cos\theta_f \\ \sum_{i=1}^m \frac{1}{r}\left(\tau_{ir} + \tau_{il}\right)\sin\theta_f \\ \sum_{i=1}^m \frac{b}{r}\left(\tau_{ir} - \tau_{il}\right) \end{bmatrix}. \tag{7.8}$$

The wheel torque vector of the combination modality is $\tau_f = [\tau_{1r}, \tau_{1l}, \ldots, \tau_{mr}, \tau_{ml}] \in \mathbf{R}^{2m}$, according to $Q_f = B_f\left(p_f\right)\tau_f$ and Eq. (7.8), the corresponding input injection matrix $B_f\left(p_f\right) \in \mathbf{R}^{3\times 2m}$ can be obtained

$$B_f\left(p_f\right) = \begin{bmatrix} \frac{1}{r}\cos\theta_f & \frac{1}{r}\cos\theta_f & \cdots & \frac{1}{r}\cos\theta_f & \frac{1}{r}\cos\theta_f \\ \frac{1}{r}\sin\theta_f & \frac{1}{r}\sin\theta_f & \cdots & \frac{1}{r}\sin\theta_f & \frac{1}{r}\sin\theta_f \\ \frac{b}{r} & -\frac{b}{r} & \cdots & \frac{b}{r} & -\frac{b}{r} \end{bmatrix}. \tag{7.9}$$

According to Eqs. (7.2), (7.4) and (7.7), the constrained kinematic equation of combination modality member can be obtained

$$M_f\left(p_f\right)\ddot{p}_f + E_f\left(p_f, \dot{p}_f\right) = B_f\left(p_f\right)\tau + A^{\mathrm{T}}\left(p_f\right)\lambda, \tag{7.10}$$

where λ is the Lagrange multiplier.

Combining the kinematic constraints, (7.10) can also be transformed without constraints to obtain the following dynamic equation

$$\bar{M}_{1f}\left(p_f\right)\dot{\eta}_f + \bar{M}_{2f}\left(p_f\right)\eta_f + \bar{E}_f\left(p_f, \dot{p}_f\right) = \bar{B}_f\left(p_f\right)\tau_f, \tag{7.11}$$

Among them, $\bar{M}_{1f}\left(p_f\right) = S^{\mathrm{T}}\left(p_f\right)M_f\left(p_f\right)S\left(p_f\right), \bar{M}_{2f}\left(p_f\right) = S^{\mathrm{T}}\left(p_f\right)M_f\left(p_f\right)\dot{S}\left(p_f\right), \bar{B}\left(p_f\right) = S^{\mathrm{T}}\left(p_f\right)B\left(p_f\right)$ and $\bar{E}_f\left(p_f, \dot{p}_f\right) = S^{\mathrm{T}}\left(p_f\right)E_f\left(p_f, \dot{p}_f\right)$.

The specific forms of the related matrices of the dynamics for the combination modality applicable to actual control scenarios can be written as

$$\begin{aligned} \bar{M}_{1f} &= \begin{bmatrix} \sum_{i=1}^m m_i & 0 \\ 0 & \sum_{i=1}^m m_i a^2 + I_{mi} \end{bmatrix}, \bar{M}_{2f} = \begin{bmatrix} 0 & 0 \\ -\sum_{i=1}^m m_i\dot{\theta}_f a & 0 \end{bmatrix}, \\ \bar{B}_f &= \begin{bmatrix} \frac{1}{r_i} & \frac{1}{r_i} & \cdots & \frac{1}{r_i} & \frac{1}{r_i}, \\ \frac{b_i}{r_i} & -\frac{b_i}{r_i} & \cdots & \frac{b_i}{r_i} & -\frac{b_i}{r_i} \end{bmatrix}, \bar{E}_f = \begin{bmatrix} \sum_{i=1}^m m_i a\dot{\theta}_f^2 \\ 0 \end{bmatrix}. \end{aligned} \tag{7.12}$$

7.2.2 Problem Description

Although reconfiguration swarm can be transformed into new modalities that adapt to current tasks and terrains through arbitrary scale combinations, they must also design combination modality controllers that match any combination scale. This chapter considers the fault-tolerant advantages of redundant actuation of combination modality and the application limitations of motion saturation, safety threats, and disturbance interference in complex environments, and designs combination modality control scheme with self-adaptive fault tolerance, safety assurance, and stable control.

Combination modality member use the kinematic and dynamic models of Eqs. (7.1) and (7.10), but the slip steering mode of the combination modality, compared to the differential steering of the modular unit, is more likely to cause kinematic uncertainties such as tire skidding. At the same time, in the dynamics of the combination modality member with redundant actuation, in addition to the lumped uncertainties such as unmodeled dynamics, identification errors, and external disturbances, the random uncertainties of the bottom combination dynamics actuation faults must also be considered due to adaptive tasks such as faults and obstacle crossing reconfiguration.

Referring to Assumption 7.1 and the kinematic uncertainty proposed by Chaw [7, 8], it is easy to obtain the kinematic model of the combination modality member under unknown sliding disturbances

$$\dot{p}_f = S_1\left(p_f\right)\eta_f + S_2\left(p_f\right)\delta_f + d_f, \tag{7.13}$$

$$A\left(p_f\right)\dot{p}_f = \delta_w\left(v_f, w_f\right), \tag{7.14}$$

where,

$$S_1\left(p_f\right) = \begin{bmatrix} \cos\theta_f & 0 \\ \sin\theta_f & 0 \\ 0 & 1 \end{bmatrix}, S_2\left(p_f\right) = \begin{bmatrix} \cos\theta_f & \sin\theta_f \\ \sin\theta_f & -\cos\theta_f \\ 0 & 0 \end{bmatrix}, d_f = \begin{bmatrix} d_{fx} \\ d_{fy} \\ d_{f\theta} \end{bmatrix}, \eta_f = \begin{bmatrix} v_f \\ w_f \end{bmatrix},$$

$$\delta_f = \begin{bmatrix} \delta_v\left(v_f, w_f\right) \\ \delta_w\left(v_f, w_f\right) \end{bmatrix}, A\left(p_f\right) = \left[\sin\theta_f, -\cos\theta_f, 0\right],$$

$\delta_v\left(v_f, w_f\right)$ and $\delta_w\left(v_f, w_f\right)$ are unknown nonlinear functions about linear velocity and angular velocity.

Assumption 7.2 *The kinematic uncertainties $\delta_v\left(v_f, w_f\right)$ and $\delta_w\left(v_f, w_f\right)$ caused by the combination modality slippage are bounded with other uncertainties d_{fx}, d_{fy}, and $d_{f\theta}$ [8, 9].*

The fault model of each actuation motor of the combination modality member is

$$\tau_f(t) = \sigma_f(t)\,u_f(t), \tag{7.15}$$

where $\tau_f = [\tau_{1r}, \tau_{1l}, \ldots, \tau_{mr}, \tau_{ml}]^T \in \mathbf{R}^{2m}$ is the actual combined vector of the torques of each wheel actuation motor of the combination modality member, $u_f = [u_{1r}, u_{1l}, \ldots, u_{mr}, u_{ml}] \in \mathbf{R}^{2m}$ is the torque distributed on average according to the lower-level control input, $\sigma_f = \text{diag}\left\{\sigma_{1r}, \sigma_{1f}, \ldots, \sigma_{mr}, \sigma_{ml}\right\} \in \mathbf{R}^{2m\times 2m}$ represents the health status of the actuation motors of the combination modality member. If $\sigma_{ir} = 1$, the right wheel motor of the ith unit in the combination modality member is in a healthy state, and $\sigma_{ir} = 0$ only indicates that the right wheel motor of the ith unit is in a loss of control fault state. The health status of other units and their actuation motors are inferred in the same way.

Remark 7.1 The types of faults in the combination modality member include various types such as failure, jamming, and loss of control. The loss of control fault indicates that the actuator has completely failed ($\sigma_{ir} = 0$ and $\sigma_{il} = 0$), while under failure and jamming faults, the actuator still has considerable actuation ability and can be treated as an unknown disturbance, which can be compensated in real time through disturbance estimation in actual applications. Therefore, this chapter considers the loss of control faults of the actuator that may cause serious losses.

The dynamic model of the combination modality member under the lower-level dynamics uncertainty where faults may occur can be expressed as follows

$$\bar{M}_{1f}\left(p_f\right)\dot{\eta}_f + \bar{M}_{2f}\left(p_f\right)\eta_f + \bar{E}_f\left(p_f, \dot{p}_f\right) = \bar{B}_f\left(p_f\right)\sigma_f u_f + \tau_{df}, \tag{7.16}$$

Here, $\tau_{df} \in \mathbf{R}^2$ represents the lumped uncertainties in the dynamics of the combination modality member, including changes in the dynamic model caused by kinematic uncertainty, unmodeled dynamics, identification errors, and unknown external disturbances.

This chapter is similar to Chaps. 4–6 in terms of kinematic saturation constraints. When designing a controller for the convergence of tracking errors in the combined modality model (7.13), it is also necessary to apply nested saturation constraints similar to Eq. (4.4) to the linear and angular velocity commands v_{fd} and w_{fd}. The saturation thresholds for the amplitude of linear and angular velocities are $\bar{v}_f$ and $\bar{w}_f$, and the saturation thresholds for their rates are $\bar{a}_f$ and $\bar{\omega}_f$. The specific nested saturation model is as follows

$$\dot{v}_{fs} = \text{sat}\left(-\bar{a}_f, \left(\text{sat}\left(-\bar{v}_f, v_{fd}, \bar{v}_f\right) - v_{fs}\right)/\tau_{fv}, \bar{a}_f\right), \tag{7.17}$$

$$\dot{w}_{fs} = \text{sat}\left(-\bar{\omega}_f, \left(\text{sat}\left(-\bar{w}_f, w_{fd}, \bar{w}_f\right) - w_{fs}\right)/\tau_{fw}, \bar{\omega}_f\right), \tag{7.18}$$

where v_{fs} and w_{fs} are the saturated linear and angular velocity control commands of the combination modality. Similarly, an adaptive auxiliary dynamic system can be used to suppress possible saturation oscillations. In the lower-level combined control model (7.16), the dynamics of the combination modality are controlled to achieve tracking of the saturated velocity commands v_{fs}, w_{fs}.

Assumption 7.3 *The norm of the dynamic uncertainty* τ_{df} *of the combination modality and its derivative* $\dot{\tau}_{df}$ *are bounded, and there exist positive numbers* $\bar{\tau}_{df}$ *and* $\bar{\tau}_{ddf}$ *such that* $\|\tau_{df}\| < \bar{\tau}_{df}$ *and* $\|\dot{\tau}_{df}\| < \bar{\tau}_{ddf}$.

Assumption 7.4 *The dynamic state* $\eta_f \in \boldsymbol{R}^2$ *of the combination modality member in the reconfiguration swarm is controllable, satisfying the redundant actuation condition of rank* $\left(B_f \sigma_f\right) = 2$.

This article aims to enhance the adaptability and safety of combination modality member in complex environment control. In the kinematic level, the position-orientation level cascade posture feedback control structure integrates safety and saturation constraints. An improved adaptive safety barrier function is introduced to ensure the safety of the combination modality member during motion. In the lower-level dynamics, considering the randomly occurring actuation faults, an adaptive fault-tolerant strategy is introduced to match the appropriate torque distribution. The AGFESO is designed to achieve smooth practical disturbance feedforward compensation, further enhancing the adaptability of the combination modality member under unknown lumped disturbances. The specific kinematics-dynamics cascaded control structure is shown in Fig. 7.2.

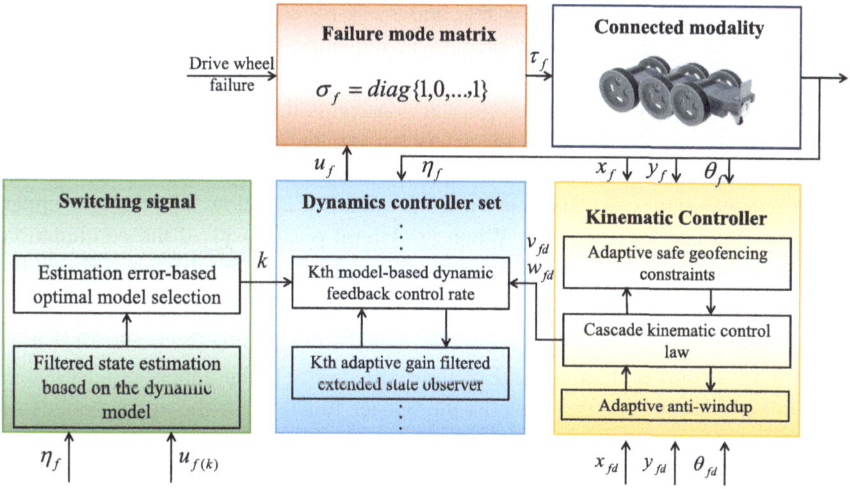

Fig. 7.2 Block diagram of the combination modality control structure

7.3 Combination Modality Robust Tracking Control in Obstacle Environment

7.3.1 Safety Barrier Avoidance Mechanism Based on Adaptive Boundary Compensation

The previous two chapters ensured the safety of the unit motion through the composite artificial potential field safety barrier. This section replaces the composite of the artificial potential field with boundary and gain adaptive adjustment means to improve the safety barrier mechanism, promoting the combination modality to smoothly avoid dangerous targets in the obstacle environment. Assume the position of the kth obstacle in the environment is $p_{ko} = [x_{ko}, y_{ko}]$. Based on the obstacle's position and the moving environment, we generate the virtual safety barriers closer to the moving boundary and the obstacle on the upper/lower/left/right side boundaries y_{fo}^{sH}, y_{fo}^{sL}, x_{fo}^{sL}, and x_{fo}^{sH}. Similarly, we generate the adaptive safety barriers on the upper/lower/left/right side boundaries y_f^H, y_f^L, x_f^L, and x_f^H that constrain the safe acceleration envelope. Similar to the safety barriers in Chap. 5, the generated barriers satisfy $|\dot{y}_{fo}^{sH}| \le \bar{\nu}_f$, $|\dot{y}_{fo}^{sL}| \le \bar{\nu}_f$, $|\dot{x}_{fo}^{sH}| \le \bar{\nu}_f$, $|\dot{x}_{fo}^{sL}| \le \bar{\nu}_f$, $|\dot{y}_f^H| \le \bar{\nu}_f$, $|\dot{y}_f^L| \le \bar{\nu}_f$, $|\dot{x}_f^H| \le \bar{\nu}_f$, $|\dot{x}_f^L| \le \bar{\nu}_f$, $|\ddot{y}_{fo}^{sH}| \le \bar{\alpha}_f$, $|\ddot{y}_{fo}^{sL}| \le \bar{\alpha}_f$, $|\ddot{x}_{fo}^{sH}| \le \bar{\alpha}_f$, $|\ddot{x}_{fo}^{sL}| \le \bar{\alpha}_f$, $|\ddot{y}_f^H| \le \bar{\alpha}_f$, $|\dot{y}_f^L| \le \bar{\alpha}_f$, $|\ddot{x}_f^H| \le \bar{\alpha}_f$, and $|\ddot{x}_f^L| \le \bar{\alpha}_f$, and $\bar{\nu}_f \le \bar{\upsilon}_f, \bar{\alpha}_f \le \bar{a}_f \in \mathbf{R}_+$. The moving assembly's upper side avoidance process in the improved adaptive safety barrier avoidance strategy is shown in Fig. 7.3. The adaptive upper side safety barrier can generate a safe acceleration envelope that constrains the combination modality to bypass the obstacle. Compared to the safety barriers in Chap. 5, the improved safety barrier also has a virtual safety barrier upper boundary closer to the obstacle. When the vertical position y_f of the combination modality is below y_f^H, the two upper boundaries deviate by $r_{sH} \in \mathbf{R}_+$. When the lateral position y_f of the combination modality crosses y_f^H, the upper boundary of the adaptive safety barrier will rapidly decrease in the opposite direction, driving the combination modality to leave the lateral position region $\left[y_f^{sH} - r_{sH}, y_f^{sH}\right]$. The specific design process is shown below, and the safety constraints in other directions are similar to the upper boundary.

The combination modality designs the barrier function for the upper bounds of the kth obstacle as follows

$$\mathcal{B}_f^{uy} = \dot{\bar{y}}_f^u - h(-\bar{y}_f^u) g_f\left(\bar{y}_f^u\right) + \left(1 - h\left(-\bar{y}_f^u\right)\right) \sqrt{\frac{\gamma_f}{\rho_f}} \bar{y}_f^u, \tag{7.19}$$

and

$$h(x) = \begin{cases} 1, & x \ge 0 \\ 0, & x < 0 \end{cases}, g_f(x) = \left(2\gamma_f \rho_f \ln\left(\cosh\left(\frac{x}{\rho_f}\right)\right)\right)^{\frac{1}{2}}, x \le 0,$$

where $\bar{y}_f^u = y_f - y_f^H$, and $\gamma_f, \rho_f \in \mathbf{R}_+$ are design gains, and $\gamma_f \le \bar{\alpha}_f$.

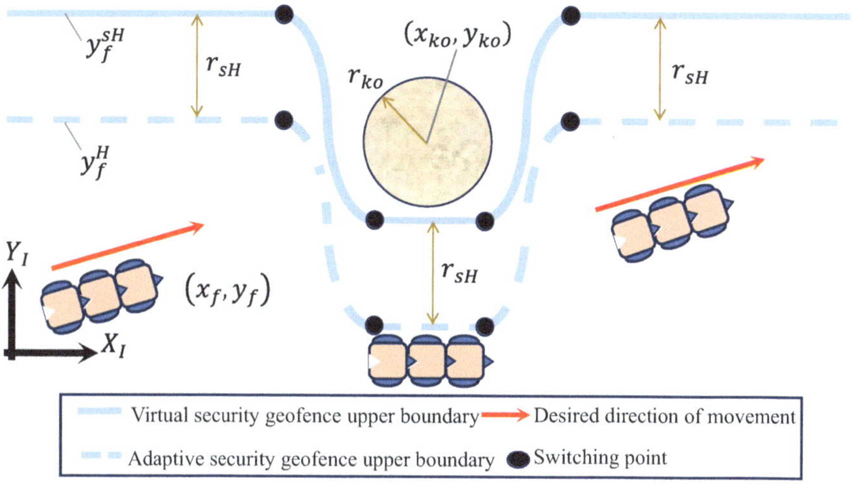

Fig. 7.3 Adaptive safe geofence settings

Taking the above upper safe geofences as an example, compared to the dynamic safe geofences in Chap. 5, an adaptive term is introduced to design and adjust y_f^{sH} to ensure that when the vertical position y_f of the combination modality member crosses $y_f^{sH} - r_{sH}$, y_f^H changes downward to make the lateral position of the combination modality member quickly leave $\left[y_f^{sH} - r_{sH}, y_f^{sH}\right]$. The specific adaptive term is shown as follows

$$p_{fo}^{uy} = \begin{cases} \frac{\varepsilon_{fH} r_{sH}}{1+e^{c_{fH}\left(y_f - y_f^H\right)}} - \frac{\varepsilon_{fH} r_{sH}}{2}, & y_f^{sH} \le y_f \le y_f^H, \\ 0, & \text{otherwise}, \end{cases} \tag{7.20}$$

where $\varepsilon_{fH}, c_{fH} \in \mathbf{R}_+$ are design gains. The variable of the improved dynamic geofence's upper safe boundary barrier function is $\bar{y}_f^u = y_f - y_f^H + p_{fo}^{uy}$. When the trajectory of the mobile combination crosses y_f^H, Eq. (7.20) can adjust y_f^H to decrease. Combined with the following acceleration envelope effect, it drives the combination modality member away from y_f^{sH}, avoiding collision. At the same time, ε_{fH} and c_{fH} need to be designed to ensure that the upper safe boundary $\bar{y}_f^u$ does not interfere with the lower safe boundary $\bar{y}_f^l$. The adaptive term for the lower safe geofence of the lateral position of the mobile combination is

$$p_{fo}^{ly} = \begin{cases} \frac{\varepsilon_{fL} r_{sL}}{1+e^{c_{fL}\left(y_f^L - y_f\right)}} - \frac{\varepsilon_{fL} r_{sL}}{2}, & y_f^L \le y_f \le y_f^{sL}, \\ 0, & \text{otherwise}, \end{cases} \tag{7.21}$$

In the formula, $r_{sL} \in \mathbf{R}_+$ is the lateral distance of $y_f^{sL} - y_f^L$, the barrier function variable is $\bar{y}_f^l = y_f^L - y_f + p_{fo}^{ly}$, and the gains ε_{fL} and $c_{fL} \in \mathbf{R}_+$ are designed to ensure that the lower safety boundary $\bar{y}^u_f$ does not interfere with the upper safety boundary $\bar{y}^l_f$. The new lateral safety barriers can be obtained as $y^u_{fo} = y^H_f - p^{uy}_{fo}$ and $y^l_{fo} = y^L_f + p^{uy}_{fo}$.

The design of the lateral upper bound safety acceleration envelope for the combination modality member relative to the obstacle is as follows

$$a^{uy}_{fo} = -\bar{\alpha}^{uy}_{fo}\mathcal{B}^{uy}_f + \mathcal{F}^{ux}_{fo}\dot{\bar{y}}^u_{fo} + \ddot{y}^u_{fo}, \tag{7.22}$$

$$a^{ly}_{fo} = -\bar{\alpha}^{ly}_{fo}\mathcal{B}^{ly}_f + \mathcal{F}^{lx}_{fo}\dot{\bar{y}}^l_{fo} - \ddot{y}^l_{fo}, \tag{7.23}$$

and

$$\mathcal{F}^{uy}_{fo} = h\left(-\bar{y}^u_f\right)\gamma_f \tanh\left(\frac{\bar{y}^u_f}{\rho_f}\right)/g_f\left(\bar{y}^u_f\right) - \left(1 - h\left(-\bar{y}^u_f\right)\right)\sqrt{\frac{\gamma_f}{\rho_f}},$$

$$\mathcal{F}^{ly}_{fo} = h\left(-\bar{y}^l_f\right)\gamma_f \tanh\left(\frac{\bar{y}^l_f}{\rho_f}\right)/g_f\left(\bar{y}^l_f\right) - \left(1 - h\left(-\bar{y}^l_f\right)\right)\sqrt{\frac{\gamma_f}{\rho_f}},$$

In the formula, $\bar{\alpha}^{ux}_{fo}$ and $\bar{\alpha}^{uy}_{fo} \in \mathbb{R}_+$ are distance-based adaptive gains, specifically as follows

$$\alpha^{uy}_{fo} = \begin{cases} \alpha^{uy}_{2fo}, & y^H_f - y_f \leq r_{2H}, \\ \alpha^{uy}_{1fo}, & \text{otherwise}, \end{cases} \tag{7.24}$$

$$\alpha^{ly}_{fo} = \begin{cases} \alpha^{ly}_{2fo}, & y_f - y^L_f \leq r_{2L}, \\ \alpha^{ly}_{1fo}, & \text{otherwise}, \end{cases} \tag{7.25}$$

$$\tau_{fo}\dot{\bar{\alpha}}^{uy}_{fo} = \bar{\alpha}^{uy}_{fo} - \alpha^{uy}_{fo}, \tag{7.26}$$

$$\tau_{fo}\dot{\bar{\alpha}}^{ly}_{fo} = \bar{\alpha}^{ly}_{fo} - \alpha^{ly}_{fo}, \tag{7.27}$$

To maintain the continuity of the gains, α^{uy}_{fo} and $\alpha^{ly}_{fo} \in \mathbf{R}_+$ are distance-adjusted gains, r_{2H} and $r_{2L} \in \mathbf{R}_+$ are triggering distances, and $\alpha^{uy}_{2fo} > \alpha^{uy}_{1fo} \in \mathbf{R}_+$ and $\alpha^{ly}_{2fo} > \alpha^{ly}_{1fo} \in \mathbf{R}_+$ are gains at different approach distances. $\bar{\alpha}^{ux}_{fo}$ and $\bar{\alpha}^{uy}_{fo}$ are the outputs of α^{ux}_{fo} and α^{uy}_{fo} after filtering, and τ_{fo} is the filter gain. The purpose of adaptive gain design is to choose a smaller gain to weaken the constraint of the safety barrier on the motion when far from the safety barrier, and to choose a larger gain to strengthen the constraint of the safety barrier on the motion trajectory within the range close to the safety barrier to ensure the safety of the combination modality member.

The safety acceleration envelopes a^{lx}_{fo} and a^{ux}_{fo} on the left and right sides are similar to the acceleration envelopes for obstacle avoidance in other directions. The final generated upper/lower side safety acceleration envelope constrains the lateral velocity command v_{fx} of the combination modality member, and the left/right side safety acceleration envelope constrains the longitudinal velocity command v_{fy} of the combination modality member, as shown below

$$\begin{cases} \dot{v}^{x}_{fs} = \operatorname{sat}\left(-\bar{\alpha}_f, \operatorname{sat}\left(-a^{lx}_{fo}, \bar{a}_{fx}, \alpha^{ux}_{fo}\right), \bar{\alpha}_f\right), \\ \dot{v}^{y}_{fs} = \operatorname{sat}\left(-\bar{\alpha}_f, \operatorname{sat}\left(-\alpha^{ly}_{fo}, \bar{a}_{fy}, \alpha^{uy}_{fo}\right), \bar{\alpha}_f\right), \end{cases} \tag{7.28}$$

and

$$\begin{aligned} &\bar{a}_{fx} = \left(v^{x}_{fs} - v_{fx}\right)/\tau_s, \bar{a}_{fy} = \left(v^{y}_{fs} - v_{fy}\right)/\tau_s, \\ &v^{x}_{fs} = \operatorname{sat}\left(-\bar{v}_f, v_{fx}, \bar{v}_f\right), v^{y}_{fs} = \operatorname{sat}\left(-\bar{v}_f, v_{fy}, \bar{v}_f\right). \end{aligned}$$

In the formula, v^{x}_{fs} and v^{y}_{fs} are the longitudinal and lateral velocity commands constrained by the safety barrier generated command, which can effectively drive the combination modality member near the obstacle to bypass the dangerous target. The saturation command and the deviation of the velocity command are $\Delta^{x}_{fs} = v^{x}_{fs} - v_{fx}$ and $\Delta^{y}_{fs} = v^{y}_{fs} - v_{fy}$.

7.3.2 Robust Anti-saturation Kinematic Controller Design with Composite Nonlinear Feedback

v_{fx} and v_{fy} are the longitudinal and lateral velocity control commands, and v_{fd} is the linear velocity control command. The corresponding relationship between the linear velocity control command and the lateral/longitudinal velocity control command is as follows

$$v_{fd} = \cos\theta_f v^{x}_{fs} + \sin\theta_f v^{y}_{fs} = \cos\left(\theta_f - \theta_{fd}\right)\sqrt{\left(v^{x}_{fs}\right)^2 + \left(v^{y}_{fs}\right)^2}, \tag{7.29}$$

Here, θ_{fd} is the desired heading angle, which can be derived from $\theta_{fd} = \arctan 2\left(v^{y}_{fs}, v^{x}_{fs}\right)$.

Combining the kinematic model in Eq. (7.13), the velocity control command mapping relationship in Eq. (7.29), and the nested motion saturation in Eqs. (7.17)–(7.18), we can obtain the kinematic dynamics under the cascaded control framework subject to nested saturation constraints as follows:

$$\begin{aligned} \dot{x}_f &= v_{fd}\cos\theta_f + \left(v_f - v_{fs}\right)\cos\theta_f + \left(v_{fs} - v_{fd}\right)\cos\theta_f + \delta_{vf}\cos\theta_f \\ &\quad + \delta_{wf}\sin\theta_f + d_{fx} \\ &= \left(\cos\theta_f v^{x}_{fs} + \sin\theta_f v^{y}_{fs}\right)\cos\theta_f + v_{fe}\cos\theta_f + \Delta_{fv}\cos\theta_f + \delta_{fx} \\ &= v^{x}_{fs} + v_{fe}\cos\theta_f + T_f\sin\theta_f + \Delta_{fv}\cos\theta_f + \delta_{fx}, \end{aligned} \tag{7.30}$$

$$\begin{aligned}\dot{y}_f &= v_{fd}\sin\theta_f + \left(v_f - v_{fs}\right)\sin\theta_f + \left(v_{fs} - v_{fd}\right)\sin\theta_f + \delta_{vf}\sin\theta_f \\ &\quad - \delta_{wf}\cos\theta_f + d_{fy} \\ &= \left(\cos\theta_f v_{fs}^x + \sin\theta_f v_{fs}^y\right)\sin\theta_f + v_{fe}\sin\theta_f + \Delta_{fv}\sin\theta_f + \delta_{fy} \\ &= v_{fs}^y + v_{fe}\sin\theta_f - T_f\cos\theta_f + \Delta_{fv}\sin\theta_f + \delta_{fy},\end{aligned} \tag{7.31}$$

$$\dot{\theta}_f = w_{fd} + \left(w_f - w_{fs}\right) + \left(w_{fs} - w_{fd}\right) + d_{f\theta} = w_{fd} + w_{fe} + \Delta_{fw} + \delta_{f\theta}, \tag{7.32}$$

where $v_{fe} = v_f - v_{fs}$ and $w_{fe} = w_f - w_{fs}$ are the velocity tracking errors caused by the lower-level kinematic control of the combination modality, $\Delta_{fv} = v_{fs} - v_{fd}$ and $\Delta_{fw} = w_{fs} - w_{fd}$ are the deviations between the saturated input and the kinematic control input, $T_f = \left(\sin\theta_{fd} - \sin\theta_f\right)v_{fs}^x - \left(\cos\theta_{fd} - \cos\theta_f\right)v_{fs}^y$, $\delta_{fx} = \delta_{vf}\cos\theta_f + \delta_{wf}\sin\theta_f + d_{fx}$, $\delta_{fy} = \delta_{vf}\sin\theta_f - \delta_{wf}\cos\theta_f + d_{fy}$, $\delta_{f\theta} = d_{f\theta}$.

From Assumption 7.2, it is known that the kinematic uncertainties δ_{fx}, δ_{fy}, and $\delta_{f\theta}$ are bounded, and there exist $\bar{\delta}_{fx}, \bar{\delta}_{fy}, \bar{\delta}_{f\theta} \in \mathbf{R}_+$, such that $|\delta_{fx}| \le \bar{\delta}_{fx}$, $|\delta_{fy}| \le \bar{\delta}_{fy}$, and $|\delta_{f\theta}| \le \bar{\delta}_{f\theta}$ hold.

Similarly, T_f in the cascaded dynamics of the combination modality satisfies the following relationship:

$$T_f = -2\sin\left(\frac{\theta_f - \theta_{fd}}{2}\right)\left\{\cos\left(\frac{\theta_f + \theta_{fd}}{2}\right)v_{fs}^x + \sin\left(\frac{\theta_f + \theta_{fd}}{2}\right)v_{fs}^y\right\}, \tag{7.33}$$

where if v_{fs}^x and v_{fs}^y are bounded, the posture loop error of the combination modality quickly converges to 0, then T_f will also quickly converge to 0.

In this section, the following composite nonlinear feedback is used to improve the full-process convergence characteristics of the adaptive auxiliary dynamic system and state feedback:

$$\rho\left(\kappa, \beta, x\right) = \kappa\left(e^{-\beta x} - e^{-\beta x(0)}\right), \tag{7.34}$$

where $\kappa, \beta \in \mathbf{R}_+$ are nonlinear gains, and $x \in \mathbf{R}$ is the feedback state.

The design of the adaptive auxiliary dynamic system under the composite nonlinear feedback for improving the kinematic safety acceleration envelope and controlling input saturation is as follows:

$$\begin{aligned}\dot{\lambda}_{fs}^x &= \cosh^2\left(\lambda_{fs}^x\right)\left(-k_{fs}^x\lambda_{fs}^x - \rho\left(\kappa_{fs}^x, \beta_{fs}^x, \lambda_{fs}^x\right)\lambda_{fs}^x + \Delta_{fs}^x\right)/\gamma_{fs}^x, \\ \dot{\lambda}_{fs}^y &= \cosh^2\left(\lambda_{fs}^y\right)\left(-k_{fs}^y\lambda_{fs}^y - \rho\left(\kappa_{fs}^y, \beta_{fs}^y, \lambda_{fs}^y\right)\lambda_{fs}^y + \Delta_{fs}^y\right)/\gamma_{fs}^y,\end{aligned} \tag{7.35}$$

$$\begin{aligned}\dot{\lambda}_{fx} &= \cosh^2\left(\lambda_{fx}\right)\left(-k_{fx}\lambda_{fx} - \rho\left(\kappa_{fx}, \beta_{fx}, \lambda_{fx}\right)\lambda_{fx} + \Delta_{fv}\cos\theta_f\right)/\gamma_{fx}, \\ \dot{\lambda}_{fy} &= \cosh^2\left(\lambda_{fy}\right)\left(-k_{fy}\lambda_{fy} - \rho\left(\kappa_{fy}, \beta_{fy}, \lambda_{fy}\right)\lambda_{fy} + \Delta_{fv}\sin\theta_f\right)/\gamma_{fy}, \\ \dot{\lambda}_{f\theta} &= \cosh^2\left(\lambda_{f\theta}\right)\left(-k_{f\theta}\lambda_{f\theta} - \rho\left(\kappa_{f\theta}, \beta_{f\theta}, \lambda_{f\theta}\right)\lambda_{f\theta} + \Delta_{fw}\right)/\gamma_{f\theta},\end{aligned} \tag{7.36}$$

where λ_{fs}^x, λ_{fs}^y, λ_{fx}, λ_{fy}, and $\lambda_{f\theta}$ are auxiliary dynamic signals, k_{fs}^x, k_{fs}^y, k_{fx}, k_{fy}, $k_{f\theta}$, γ_{fs}^x, γ_{fs}^y, γ_{fx}, γ_{fy}, and $\gamma_{f\theta}$ are positive linear adaptive gains, κ_{fs}^x, κ_{fs}^y, κ_{fx}, κ_{fy}, $\kappa_{f\theta}$, β_{fs}^x, β_{fs}^y, β_{fx}, β_{fy}, and $\beta_{f\theta}$ are positive nonlinear design gains.

The longitudinal and lateral position tracking errors of the combination modality member at the kinematic level are defined as:

$$x_{fe} = x_f - x_{fd} - \gamma_{fx} \tanh \lambda_{fx} - \gamma_{fs}^x \tanh \lambda_{fs}^x, \tag{7.37}$$

$$y_{fe} = y_f - y_{fd} - \gamma_{fy} \tanh \lambda_{fy} - \gamma_{fs}^y \tanh \lambda_{fs}^y, \tag{7.38}$$

The longitudinal and lateral position control laws for the combination modality at the kinematic level are designed as:

$$\begin{aligned} v_{fx} = &-\bar{k}_{fx} \tanh\left(\frac{x_{fe}}{\bar{k}_{fx}}\right) - \rho\left(\bar{\kappa}_{fx}, \bar{\beta}_{fx}, x_{fe}\right) \tanh\left(\frac{x_{fe}}{\bar{k}_{fx}}\right) - k_{fx}\lambda_{fx} - k_{fs}^x \lambda_{fs}^x + \dot{x}_{fd} \\ &- \rho\left(\kappa_{fx}, \beta_{fx}, \lambda_{fx}\right) \lambda_{fx} - \rho\left(\kappa_{fs}^x, \beta_{fs}^x, \lambda_{fs}^x\right) \lambda_{fs}^x - \epsilon_{fx} \tanh\left(\frac{x_{fe}}{\mu_{fx}}\right), \end{aligned} \tag{7.39}$$

$$\begin{aligned} v_{fy} = &-\bar{k}_{fy} \tanh\left(\frac{y_{fe}}{\bar{k}_{fy}}\right) - \rho\left(\bar{\kappa}_{fy}, \bar{\beta}_{fy}, y_{fe}\right) \tanh\left(\frac{y_{fe}}{\bar{k}_{fy}}\right) - k_{fy}\lambda_{fy} - k_{fs}^y \lambda_{fs}^y + \dot{y}_{fd} \\ &- \rho\left(\kappa_{fy}, \beta_{fy}, \lambda_{fy}\right) \lambda_{fy} - \rho\left(\kappa_{fs}^y, \beta_{fs}^y, \lambda_{fs}^y\right) \lambda_{fs}^y - \epsilon_{fy} \tanh\left(\frac{y_{fe}}{\mu_{fy}}\right), \end{aligned} \tag{7.40}$$

Among them, $\bar{k}_{fx}$, $\bar{k}_{fy}$, $\bar{\kappa}_{fx}$, $\bar{\kappa}_{fy}$, $\bar{\beta}_{fx}$, and $\bar{\beta}_{fy} \in \mathbf{R}_+$ are the controller design gains, $\epsilon_{fx}, \epsilon_{fy}, \mu_{fx}$, and $\mu_{fy} \in \mathbf{R}_+$ are the switching term gains, and they satisfy $\epsilon_{fx} \geq \bar{\delta}_{fx}$, $\epsilon_{fy} \geq \bar{\delta}_{fy}$.

The attitude error is defined as $\theta_{fe} = \theta_f - \theta_{fd} - \gamma_{f\theta} \tanh \lambda_{f\theta}$, and the attitude control law designed to stabilize the attitude tracking error is

$$\begin{aligned} w_{fd} = &-\bar{k}_{f\theta}\theta_{fe} + \dot{\theta}_{fd} - \rho\left(\bar{\kappa}_{f\theta}, \bar{\beta}_{f\theta}, \theta_{fe}\right)\theta_{fe} - k_{f\theta}\lambda_{f\theta} \\ &- \rho\left(\kappa_{f\theta}, \beta_{f\theta}, \lambda_{f\theta}\right)\lambda_{f\theta} - \epsilon_{f\theta} \tanh\left(\frac{\theta_{fe}}{\mu_{f\theta}}\right), \end{aligned} \tag{7.41}$$

where $\bar{k}_{f\theta}$, $\bar{\kappa}_{f\theta}$, and $\bar{\beta}_{f\theta} \in \mathbf{R}_+$ are the controller design gains, $\epsilon_{f\theta}$ and $\mu_{f\theta} \in \mathbf{R}_+$ are the switching term gains, and they satisfy $\epsilon_{f\theta} \geq \bar{\delta}_{f\theta}$. The robust switching terms in Eqs. (7.39)–(7.41) are designed to suppress the uncertainties in the dynamics.

7.4 Combined Modality Dynamic Fault-Tolerant Control Under Actuator Failures

7.4.1 Smooth Disturbance Estimation with Noise Oscillation and Peaking Phenomenon Rejection

The combined modality, compared to a single unit, operates in a more complex and harsh environment, thus requiring more precise disturbance estimation and compensation to enhance the control robustness of the combined modality formation in complex environments. The improvement of disturbance estimation performance depends on the high gain of the disturbance observer. However, high gain can easily trigger two adverse conditions: the peaking phenomenon caused by the initial observer error and the high-frequency oscillation caused by the noise in the measurement signal, leading to a complete loss of control of the combined modality under the feedforward compensation control structure. Different from the peaking phenomenon suppression strategies of the fourth chapter ASESO and the sixth chapter MESO fusion estimation, this section proposes an initial small, steady-state high adaptive gain to avoid the undesired peaking phenomenon in the initial stage and the steady-state status/disturbance estimation accuracy. However, high gain can easily trigger high-frequency oscillations of the state and disturbance estimation signals in the noisy output, so a low-pass filter is introduced in the measurement output to smooth the state signals of the combined modality member and suppress the high-frequency noise dynamics. In summary, by integrating the dynamic changes of adaptive gain and low-pass filtering means, an AGFESO that can simultaneously achieve peaking phenomenon and noise oscillation suppression is designed, and the obtained high-precision smooth disturbance estimation signal is introduced into the bottom dynamics of the combination modality through a feedforward compensation method to enhance the robustness of velocity control.

Under the premise of satisfying the redundant allocation of Assumption 7.4, there exist N kinds of fault situations in the combined modality member, $\sigma_{f(k)}$ and $u_{f(k)}$, $k = 1, \ldots, N$, are the health status matrix and the predetermined allocation torques of each wheel under the kth situation. From the combined modality dynamics (7.16), the control model under the kth fault state can be obtained

$$\dot{\eta}_f = \bar{M}_{1f}^{-1}\left(\bar{B}_f\left(p_f\right)\sigma_{f(k)}u_f + \tau_{df} - \bar{M}_{2f}\left(p_f\right)\eta_f - \bar{E}_f\left(p_f, \dot{p}_f\right)\right). \tag{7.42}$$

Let $x_{1f} = \left[v_f, w_f\right]^{\mathrm{T}}$, $x_{2f} = \bar{M}_{1f}^{-1}\tau_{df}$, Eq. (7.42) can be rewritten as

$$\begin{aligned} \dot{x}_{1f} &= \bar{M}_{1f}^{-1}\bar{B}_f\left(p_f\right)\sigma_{f(k)}u_f + F_f\left(\cdot\right) + \bar{M}_{1f}^{-1}\tau_{df}, \\ \dot{x}_{2f} &= h_f, \end{aligned} \tag{7.43}$$

where $h_f = \bar{M}_{1f}^{-1} \frac{d\tau_{df}}{dt}$ is the lumped disturbances including internal model uncertainty and external random disturbance, $F_f(\cdot) = \bar{M}_{1f}^{-1} \left(-\bar{M}_{2f}(p_f)\eta_f - \bar{E}_f(p_f, \dot{p}_f)\right)$ is the measurable model nonlinearity, and $\sigma_{f(k)}$ is the allocation matrix of the kth fault state. Let $y_f = x_{1f} + \xi_f \in \mathbf{R}^2$ be the noisy measurable output, where $\xi_f \in \mathbf{R}^2$ is the random noise. The design of the adaptive gain filter extented state observer is as follows

$$\begin{cases} \dot{\hat{x}}_{1f} = \hat{x}_{2f} + \bar{M}_{1f}^{-1}\bar{B}_f(p_f)\sigma_{f(k)}u_f + F_f(\cdot) + L_1(x_p - \hat{x}_{1f}), \\ \dot{\hat{x}}_{2f} = L_2(x_p - \hat{x}_{1f}), \\ \dot{x}_p = -A_f^{-1}(x_p - y_f), \end{cases} \tag{7.44}$$

where $\hat{x}_{1f}, \hat{x}_{2f} \in \mathbf{R}^2$ are the estimates of velocity and lumped uncertainty respectively, $A_f = \text{diag}\{\tau_{pv}, \tau_{pw}\} \in \mathbf{R}_+^2$ is the filter gain matrix, $\tau_{pv}, \tau_{pw} \in \mathbf{R}_+$ are the filter design parameters, $L_1 = \text{diag}\{l_{1v}, l_{1w}\}, L_2 = \text{diag}\{l_{2v}, l_{2w}\} \in \mathbf{R}_+^2$ are the adaptive gain matrices for disturbance estimation, and the corresponding observer gain adaptive law is

$$\begin{cases} l_{1v} = 2\omega_{0v}, l_{2v} = \omega_{0v}^2, \omega_{0v}(0) = \omega_{1v}, \\ \dot{\omega}_{0v} = (\omega_{2v} + \omega_{2v}\text{sgn}(t - T_\omega) + \omega_{1v} - \omega_{0v})/\tau_\omega, \end{cases} \tag{7.45}$$

$$\begin{cases} l_{1w} = 2\omega_{0w}, l_{2w} = \omega_{0w}^2, \omega_{0w}(0) = \omega_{1w}, \\ \dot{\omega}_{0w} = (\omega_{2w} + \omega_{2w}\text{sgn}(t - T_\omega) + \omega_{1w} - \omega_{0w})/\tau_\omega, \end{cases} \tag{7.46}$$

where $\omega_{0v}, \omega_{0w} \in \mathbf{R}_+$ are the dynamically changing adaptive gains, $\omega_{1v}, \omega_{1w} \in \mathbf{R}_+$ are the initial stage observer gains, $\omega_{2v}, \omega_{2w} \in \mathbf{R}_+$ are the additional observer gains in the steady state, T_ω and τ_ω are the gain switching time and adaptive coefficient, respectively. The adaptive laws (7.45)–(7.46) can ensure a smooth transition between the two stages, the low gain in the initial stage can suppress the peaking phenomenon caused by the initial error, and the high gain after the smooth transition can ensure high precision disturbance estimation.

Define the disturbance estimation error vector as $\tilde{x}_f = \left[\tilde{x}_{1f}, \tilde{x}_{2f}, \tilde{x}_p\right]^{\mathrm{T}} \in \mathbf{R}^6$, where $\tilde{x}_{1f} = x_{1f} - \hat{x}_{1f}$, $\tilde{x}_{2f} = x_{2f} - \hat{x}_{2f}$ and $\tilde{x}_p = x_{1f} - x_p$. The scaled estimation error is $\tilde{e}_{fo} = H_\omega \tilde{x}_f \in \mathbf{R}^6$, where $H_\omega = \text{diag}\{\omega_{0v}, \omega_{0w}, 1, 1, 1, 1\}$. The error dynamics of the disturbance estimation can be obtained from Eqs. (7.43) and (7.44)

$$\bar{L}_1\dot{e}_{fo} = \bar{L}_1 H_\omega L_{fo} H_\omega^{-1} e_{fo} + \bar{L}_1 D_{fo} h_f + \bar{L}_1 E_{fo}\left(\xi_f + \dot{x}_{1f}\right), \tag{7.47}$$

where

$$L_{fo} = \begin{bmatrix} -L_1 & \mathbf{I}_2 & -L_1 \\ -L_2 & \mathbf{0}_2 & -L_2 \\ \mathbf{0}_2 & \mathbf{0}_2 & -A_f \end{bmatrix}, D_{fo} = \begin{bmatrix} \mathbf{0}_2 \\ \mathbf{I}_2 \\ \mathbf{0}_2 \end{bmatrix}, E_{fo} = \begin{bmatrix} \mathbf{0}_2 \\ \mathbf{0}_2 \\ \mathbf{I}_2 \end{bmatrix}, \bar{L}_1 = \begin{bmatrix} L_1^{-1} & \mathbf{0}_2 & \mathbf{0}_2 \\ \mathbf{0}_2 & L_1^{-1} & \mathbf{0}_2 \\ \mathbf{0}_2 & \mathbf{0}_2 & L_1^{-1} \end{bmatrix},$$

where $\mathbf{I}_2 = \text{diag}\{1, 1\}$, $\mathbf{0}_2 = \text{diag}\{0, 0\}$, $L_{fo} \in \mathbf{R}^6$ is Hurwitz, then $\bar{L}_{fo} = \bar{L}_1 H_\omega L_{fo} H_\omega^{-1} \in \mathbf{R}^6$ is also Hurwitz, thus there exist positive definite matrices $P_{fo}, Q_{fo} \in \mathbf{R}^6$ such that $\bar{L}_{fo}^{\text{T}} P_{fo} + P_{fo}\bar{L}_{fo} = -2\mathbf{I}_6 = \text{diag}\{1, \ldots, 1\} \in \mathbf{R}^6$.

Remark 7.2 Theoretically, this section can extract smooth state signals from high-frequency noise output through low-pass filtering, where $\xi_f \approx 0$, $\dot{x}_{1f}$ can be set as a bounded signal referring to the method of proving the stability of dynamic surface control.

7.4.2 Adaptive Fault-Tolerant and Anti-disturbance Dynamic Controller Design

For each fault situation $\sigma_{f(k)}$, the dynamic controller design that integrates the dynamic model and disturbance estimation feedforward compensation is

$$u_{f(k)} = \left(\bar{M}_{1f}^{-1} \bar{B}_{1f} \sigma_{f(k)}\right)^+ \left(-K_{f\eta}\eta_{fe} + \dot{\eta}_{fd} - F_f(\cdot) - \hat{x}_{2f}\right), \tag{7.48}$$

where $K_{f\eta} = \text{diag}\{k_{fv}, k_{fw}\} \in \mathbf{R}_+^2$ is the diagonal gain matrix, $\eta_{fe} = \eta_f - \eta_{fd}$ is the velocity tracking error, where the desired signal of the bottom velocity dynamics under the cascaded structure, $\eta_{fd} = \left[v_{fs}, w_{fs}\right]^{\text{T}}$, is affected by the nested saturation kinematic control input.

From Eq. (7.16), the following model can be obtained

$$\dot{\eta}_f = \bar{M}_{1f}^{-1} \bar{B}_{1f} \sigma_f u_f + F_f(\cdot) + \bar{M}_{1f}^{-1} \tau_{df}, \tag{7.49}$$

From Eqs. (7.44), (7.48), and (7.49), and assuming that $\sigma_{f(k)}$ in the controller is consistent with the actual σ_f, the dynamic error of the combination modality dynamics can be obtained as

$$\dot{\eta}_{fe} = \bar{M}_{1f}^{-1} \bar{B}_f \sigma_f u_{f(k)} + F_f(\cdot) + x_{2f} - \dot{\eta}_{fd} = -K_{f\eta}\eta_{fe} + \tilde{x}_{2f}. \tag{7.50}$$

In actual applications, the health status matrix $\sigma_{f(k)}$ of the combination modality is unknown. If the health status matrix in the dynamic controller (7.48) is inconsistent with the actual control model (7.16), it will lead to a degradation of the overall control performance or even loss of control. Therefore, this section uses the measured velocity and the known dynamic model parameters in Eq. (7.16), matches them with the finite elements of the health status matrix $\sigma_{f(k)}$ (the total number of elements in the set is N), obtains the velocity state and its estimation error under full health and each fault state based on the model estimation, and takes the minimum estimation error as the optimization goal to obtain the torque distribution that best fits the actual health status of the combination modality.

The disturbance estimation and feedforward compensation strategy in this book can largely eliminate the influence of $\bar{M}_{1f}^{-1}\tau_{df}$ on the estimation error, and the influence of measurement noise on the estimation error can be improved by filtering means to enhance the reliability of matching [10]. Therefore, the dynamic equation used for filtering estimation and its equation after low-pass filtering are shown as follows

$$\dot{\eta}_f = \bar{M}_{1f}^{-1}\bar{B}_{1f}\sigma_f\bar{u}_f + F_f(\cdot), \tag{7.51}$$

$$\frac{s}{s+\gamma}\left[\eta_f\right] = \frac{1}{s+\gamma}\left[\bar{M}_{1f}^{-1}\bar{B}_{1f}\sigma_f\bar{u}_f\right] + \frac{1}{s+\gamma}\left[F_f(\cdot)\right], \tag{7.52}$$

where $\bar{u}_f = \left(\bar{M}_{1f}^{-1}\bar{B}_{1f}\sigma_{f(k)}\right)^{+}\left(-K_{f\eta}\eta_{fe} + \dot{\eta}_{fd} - F_f(\cdot)\right)$, $\dot{\eta}_f = s\left[\eta_f\right]$, $\frac{1}{s+\gamma}[\cdot]$ is the output after $\cdot$ is filtered, and $\gamma \in \mathbf{R}_+$ is the filter gain.

According to Eq. (7.52), the velocity of the combination modality estimated based on the dynamic model can be obtained as

$$\eta_f = \frac{\gamma}{s+\gamma}\left[\eta_f\right] + \frac{1}{s+\gamma}\left[\bar{M}_{1f}^{-1}\bar{B}_{1f}\sigma_f\bar{u}_f\right] + \frac{1}{s+\gamma}\left[F_f(\cdot)\right]. \tag{7.53}$$

The estimated velocity of the combination modality under the fault state $\sigma_{f(k)}$, $k \in \{1, \ldots, N\}$ based on the dynamic model is

$$\hat{\eta}_{f(k)} = \frac{\gamma}{s+\gamma}\left[\eta_f\right] + \frac{1}{s+\gamma}\left[\bar{M}_{1f}^{-1}\bar{B}_{1f}\sigma_{f(k)}\bar{u}_{f(k)}\right] + \frac{1}{s+\gamma}\left[F_f(\cdot)\right], \tag{7.54}$$

where $\bar{u}_{f(k)} = \left(\bar{M}_{1f}^{-1}\bar{B}_{1f}\sigma_{f(k)}\right)^{+}\left(-K_{f\eta}\eta_{fe} + \dot{\eta}_{fd} - F_f(\cdot)\right)$.

Combining Eqs. (7.53) and (7.54), the estimated error and its corresponding cost function can be obtained as

$$\tilde{\eta}_{f(k)} = \eta_f - \hat{\eta}_{f(k)}, \tag{7.55}$$

$$J_{f(k)} = \tilde{\eta}_{f(k)}^{\mathrm{T}}\tilde{\eta}_{f(k)}, k \in \{1, \ldots, N\}. \tag{7.56}$$

According to the cost function, select the health status matrix that best matches the actual driving state of the combination modality, that is,

$$k = \arg\min_{k=1,\ldots,N} J_{f(k)}, \tag{7.57}$$

Based on the optimal match selected, the corresponding controller $u_{f(k)}$ can choose the most suitable wheelbase match $\sigma_{f(k)}$.

Remark 7.3 Among the N fault states, it can be divided into $\sigma_{f(a)}$ that matches the current fault state and $\sigma_{f(b)}, b \in \{1, \ldots, N, b \neq a\}$ that does not match the current fault state. The cost function of the matching fault state $J_{f(a)} = 0$, and the cost function of the non-matching fault state $J_{f(b)} > 0$. Therefore, it is feasible to choose the most suitable current fault state from a finite combination of fault states. In the controller of Eq. (7.49), the health status matrix can choose $\sigma_{f(a)}$, to avoid performance degradation and loss of control of the combination modality due to motor faults during the task.

7.5 Cascade Stability Analysis

This section analyzes the overall stability of the combination modality from the perspectives of kinematics and dynamics. The desired input at the kinematic level is the linear velocity and angular velocity commands with nested saturation constraints, and both the desired input and its derivatives are bounded. The purpose of proving dynamic stability is to verify that the velocity tracking error of the combination modality will converge to a sufficiently small error range under the action of the controller. The overall proof idea of this chapter refers to Lashkari's hierarchical proof of the stability of kinematics and dynamics of modular units [1]. In practical applications, dynamics has faster response and control bandwidth than kinematics. Although it is possible to equate the velocity command to the actual velocity during controller design and application, the interference effect of the bottom dynamics on the upper kinematics still needs to be considered. This section considers the control error at the dynamic level as the uncertainty at the kinematic level and analyzes its impact on the precision of the upper kinematic level trajectory control. The following is the specific process of stability analysis of the combination modality member.

Theorem 7.1 *For the combination modality dynamics model (7.16) that may fail, and the dynamics controller of the combination modality matches the actual fault state,* $\sigma_f = \sigma_{f(k)}$*, satisfying Assumptions 7.3–7.4, the combination modality dynamics velocity tracking controller (7.48) designed with the fusion AGFESO disturbance feedforward compensation (7.44) makes the closed-loop system signal semi-global uniformly bounded, and the velocity tracking error converges.*

Proof Choose the following Lyapunov function under certain driving conditions at the dynamic level of the combination modality

$$V_{1f} = \frac{1}{2}\eta_{fe}^{\mathrm{T}}\eta_{fe} + \frac{1}{2}e_{fo}^{\mathrm{T}}P_{fo}e_{fo}. \tag{7.58}$$

Define the sets $\Omega_{1f} = \{(\eta_{fd}, \dot{\eta}_{fd}) : \eta_{fd}^{\mathrm{T}}\eta_{fd} + \dot{\eta}_{fd}^{\mathrm{T}}\dot{\eta}_{fd} \leq p_\eta\}$ and $\Omega_{2f} = \{\eta_{fe}^{\mathrm{T}}\eta_{fe} + e_{fo}^{\mathrm{T}}e_{fo} \leq 2p\}$, where $\Omega_{1f} \in \mathbf{R}^6$, $\Omega_{2f} \in \mathbf{R}^8$, and thus $\Omega_{1f} \times \Omega_{2f} \in \mathbf{R}^{14}$. Simultaneously, according to Eq. (7.47), by defining $G_{f(k)}$, we can obtain $\dot{e}_{fo} = \bar{L}_1^{-1}\bar{L}_{fo}e_{fo} + D_{fo}h_f + G_{f(k)}\left(\eta_{fe}, e_{fo}, \eta_{fd}, \dot{\eta}_{fd}\right)$.

Differentiating equation (7.58), we get

$$\dot{V}_{1f} \leq -\eta_{fe}^{\mathrm{T}} K_{f\eta}\eta_{fe} + \eta_{fe}^{\mathrm{T}}\tilde{x}_{2f} - \lambda_{\min}\left(\bar{L}_1^{-1}\right) e_{fo}^{\mathrm{T}} e_{fo} + e_{fo}^{\mathrm{T}} D_{fo} h_f + e_{fo}^{\mathrm{T}} G_{f(k)}. \tag{7.59}$$

For $V_{1f}(0) \leq p$, where p is a positive real number. When $V_{1f} = p$, $\|G_{f(k)}\|$ is bounded, and the upper bound of $\|G_{f(k)}\|$ in the set $\Omega_{1f} \times \Omega_{2f}$ is defined as $M_{f(k)}$, then $\|G_{f(k)}\|^2/M_{f(k)}^2 - 1 \leq 0$. We will prove that when $V_{1f} \leq p$, $\dot{V}_{1f} \leq 0$.

Through Young's inequality, $\dot{V}_{1f}$ can be further simplified as

$$\begin{aligned}\dot{V}_{1f} \leq & -\eta_{fe}^{\mathrm{T}} K_{f\eta}\eta_{fe} - \lambda_{\min}\left(\bar{L}_1^{-1}\right) e_{fo}^{\mathrm{T}} e_{fo} + \frac{1}{2}\|\eta_{fe}\|^2 + \frac{1}{2}\|e_{fo}\|^2 + \frac{1}{2}\|e_{fo}\|^2 \\ & + \frac{1}{2}\bar{h}_f^2 + \frac{1}{2}\|e_{fo}\|^2 G_{f(k)}^2 + \frac{1}{2}.\end{aligned} \tag{7.60}$$

Choose parameters

$$\lambda_{\min}\left(K_{f\eta}\right) \geq \frac{1}{2} + r_{1f}, \ \frac{\lambda_{\min}\left(\bar{L}_1^{-1}\right)}{\lambda_{\max}\left(P_{fo}\right)} \geq 1 + \frac{1}{2}M_{f(k)}^2 + r_{1f},$$

where r_{1f} is a positive number to be designed.

Then we have

$$\dot{V}_{1f} \leq -2r_{1f}V_{1f} + Q_{1f} + \left(\frac{\|G_{f(k)}\|^2}{M_{f(k)}^2} - 1\right)\frac{\|e_{fo}\|^2}{2}, \tag{7.61}$$

where $Q_{1f} = \frac{1}{2}\bar{h}_f^2 + \frac{1}{2}$.

From Eq. (7.61), we know that if we choose $r_{1f} \geq \frac{Q_{1f}}{2p}$, then $\dot{V}_{1f} \leq -2r_{1f}p + Q_{1f}$. Since when $V_{1f} = p$, $\|G_{f(k)}\| \leq M_{f(k)}$ holds, so when $V_{1f} = p$, $\dot{V}_{1f} \leq -r_{1f}V_{1f} + Q_{1f}$. Therefore, $V_{1f} \leq p$ is an invariant set, i.e., if $V_{1f}(0) \leq p$, then for all $t > 0$, $V_{1f} \leq p$. Given the condition $V_{1f}(0) \leq p$, we can deduce

$$\dot{V}_{1f} \leq -2r_{1f}V_{1f} + Q_{1f}. \tag{7.62}$$

By using the inequality solving theorem, we can easily get

$$V_{1f} \leq \frac{Q_{1f}}{2r_{1f}} + \left(V_{1f}(0) - \frac{Q_{1f}}{2r_{1f}}\right)e^{-2r_{1f}t}. \tag{7.63}$$

In summary, by increasing the dynamic control gain, disturbance estimation gain, and filter gain, r_{1f} can be increased. Therefore, the dynamic control parameters can be adjusted to make the tracking error meet the requirements of the kinematic response. □

The combination modality kinematic controller and the dynamic controller are cascaded, and the stability of the kinematic layer controller that integrates the dynamic velocity tracking error is demonstrated as follows.

Theorem 7.2 *For the combination modality kinematic model (7.13), satisfying Assumption 7.2, and the velocity command is subject to nested saturation constraints (7.17) and the safety acceleration envelope constraints generated by the adaptive safety barrier (7.28), the position-heading cascaded composite nonlinear feedback robust kinematic controller (7.39), (7.40), (7.41) is designed by integrating the adaptive auxiliary dynamic system (7.35)–(7.36). This makes the combination modality pose tracking errors x_{fe}, y_{fe}, and θ_{fe} stable, bounded, and regulated.*

Proof Let $\xi_{fx} = \tanh\left(\frac{x_{fe}}{k_{fx}}\right)$ and $\xi_{fy} = \tanh\left(\frac{y_{fe}}{k_{fy}}\right)$, from Eqs. (7.30) and (7.39), we know that the dynamics of the kinematic layer position and pose errors are

$$\begin{aligned}\dot{x}_{fe} &= -\bar{k}_{fx}\xi_{fx} - \rho\left(\bar{\kappa}_{fx}, \bar{\beta}_{fx}, x_{fe}\right)\xi_{fx} - \epsilon_{fx}\tanh\left(\frac{x_{fe}}{\mu_{fx}}\right) + v_{fe}\cos\theta_f \\ &\quad + T_f\sin\theta_f + \delta_{fx}, \qquad (7.64)\\ \dot{y}_{fe} &= -\bar{k}_{fy}\xi_{fy} - \rho\left(\bar{\kappa}_{fy}, \bar{\beta}_{fy}, y_{fe}\right)\xi_{fy} - \epsilon_{fy}\tanh\left(\frac{y_{fe}}{\mu_{fy}}\right) + v_{fe}\sin\theta_f \\ &\quad - T_f\cos\theta_f + \delta_{fy}, \qquad (7.65)\\ \dot{\theta}_{fe} &= -\bar{k}_{f\theta}\theta_{fe} - \rho\left(\bar{\kappa}_{f\theta}, \bar{\beta}_{f\theta}, \theta_{fe}\right)\theta_{fe} - \epsilon_{f\theta}\tanh\left(\frac{\theta_{fe}}{\mu_{f\theta}}\right) + w_{fe} + \delta_{f\theta}, \qquad (7.66)\end{aligned}$$

Define the Lyapunov function of the combination modality kinematic layer relative to the position and pose errors as

$$V_{2f} = \bar{k}_{fx}\ln\left\{\cosh\left(\frac{x_{fe}}{k_{fx}}\right)\right\} + \bar{k}_{fy}\ln\left\{\cosh\left(\frac{y_{fe}}{k_{fy}}\right)\right\} + \frac{1}{2}\theta_{fe}^2. \qquad (7.67)$$

Differentiating V_{2f} gives

$$\begin{aligned}\dot{V}_{2f} &= -\bar{k}_{fx}\xi_{fx}^2 - \bar{k}_{fy}\xi_{fy}^2 - \rho\left(\bar{\kappa}_{fx}, \bar{\beta}_{fx}, x_{fe}\right)\xi_{fx}^2 - \rho\left(\bar{\kappa}_{fy}, \bar{\beta}_{fy}, y_{fe}\right)\xi_{fy}^2 \\ &\quad - \bar{k}_{f\theta}\theta_{fe}^2 - \rho\left(\bar{\kappa}_{f\theta}, \bar{\beta}_{f\theta}, \theta_{fe}\right)\theta_{fe}^2 - \epsilon_{fx}\tanh\left(\frac{x_{fe}}{\mu_{fx}}\right)\xi_{fx} \\ &\quad - \epsilon_{fy}\tanh\left(\frac{y_{fe}}{\mu_{fy}}\right)\xi_{fy} - \epsilon_{f\theta}\tanh\left(\frac{\theta_{fe}}{\mu_{f\theta}}\right)\theta_{fe} + \xi_{fx}v_{fe}\cos\theta_f \\ &\quad + \xi_{fx}T_f\sin\theta_f + \xi_{fy}v_{fe}\sin\theta_f - \xi_{fy}T_f\cos\theta_f + \theta_{fe}w_{fe} \\ &\quad + \xi_{fx}\delta_{fx} + \xi_{fy}\delta_{fy} + \theta_{fe}\delta_{f\theta}. \qquad (7.68)\end{aligned}$$

Assumption 7.5 *Let $\mu > 0$. For any $\chi \in \boldsymbol{R}$, there exists a constant $\mu > 0$ such that the following inequality holds*

$$0 \leq |\chi| - \chi \tanh\left(\frac{\chi}{\mu}\right) < \iota\mu, \iota = 0.2785. \tag{7.69}$$

According to Theorem 6.2, v_{fe} and w_{fe} have upper bounds, making $|v_{fe}| < \bar{v}_{fe} \in \mathbf{R}_+$ and $|w_{fe}| < \bar{w}_{fe} \in \mathbf{R}_+$ hold. The boundaries $\bar{v}_{fe}$ and $\bar{w}_{fe}$ can be adjusted through the dynamics-related gain.

Combining Assumption 7.5 and Young's inequality, $\dot{V}_{2f}$ satisfies

$$\begin{aligned}
\dot{V}_{2f} \leq &- \left(\bar{k}_{fx} + \rho\left(\bar{\kappa}_{fx}, \bar{\beta}_{fx}, x_{fe}\right)\right)\xi_{fx}^2 - \left(\bar{k}_{fy} + \rho\left(\bar{\kappa}_{fy}, \bar{\beta}_{fy}, y_{fe}\right)\right)\xi_{fy}^2 \\
&- \left(\bar{k}_{f\theta} + \rho\left(\bar{\kappa}_{f\theta}, \bar{\beta}_{f\theta}, \theta_{fe}\right)\right)\theta_{fe}^2 + \iota\epsilon_{fx}\mu_{fx} + \iota\epsilon_{fy}\mu_{fy} + \iota\epsilon_{f\theta}\mu_{f\theta} \\
&+ \frac{3}{4}\xi_{fx}^2 + \frac{3}{4}\xi_{fx}^2 + \frac{1}{4}\theta_{fe}^2 + \bar{v}_{fe}^2 + \bar{w}_{fe}^2 + T_{fx}^2 + T_{fy}^2,
\end{aligned} \tag{7.70}$$

where $T_{fx} = T_f \cos\theta_f$ and $T_{fy} = T_f \sin\theta_f$.

Choose parameters

$$\begin{aligned}
&\bar{k}_{fx} + \rho\left(\bar{\kappa}_{fx}, \bar{\beta}_{fx}, x_{fe}\right) \geq r_{2fx} + \frac{3}{4}, \bar{k}_{fy} + \rho\left(\bar{\kappa}_{fy}, \bar{\beta}_{fy}, y_{fe}\right) \geq r_{2fy} + \frac{3}{4}, \\
&\bar{k}_{f\theta} + \rho\left(\bar{\kappa}_{f\theta}, \bar{\beta}_{f\theta}, \theta_{fe}\right) \geq r_{2f\theta} + \frac{1}{4},
\end{aligned}$$

where r_{2fx}, r_{2fy}, and $r_{2f\theta}$ are positive numbers to be designed.

Therefore,

$$\dot{V}_{2f} \leq -r_{2fx}\xi_{fx}^2 - r_{2fy}\xi_{fy}^2 - r_{2f\theta}\theta_{fe}^2 + Q_{2f}, \tag{7.71}$$

where $Q_{2f} = \iota\epsilon_{fx}\mu_{fx} + \iota\epsilon_{fy}\mu_{fy} + \iota\epsilon_{f\theta}\mu_{f\theta} + \bar{v}_{fe}^2 + \bar{w}_{fe}^2 + T_{fx}^2 + T_{fy}^2$, Q_{2f} is a bounded signal. Combining Eqs. (7.67) and (7.71), the position tracking error of the combination modality kinematic level is ultimately bounded, and the precision of the steady-state error can be adjusted through the kinematic control gain. □

Remark 7.4 $v_{fe}\cos\theta_f$, $v_{fe}\sin\theta_f$, and w_{fe} are the velocity errors of the kinematic level position and attitude loop, which act as disturbances in the position and attitude loop. $T_f \sin\theta_f$ and $T_f \cos\theta_f$ are the control errors of the attitude loop level position loop, which also act as disturbances in the position loop. In the multi-loop cascade control framework, these cascade control errors are quickly converged through the inner loop, that is, the inner loop obtains a higher control precision of the outer loop through a larger control gain relative to the outer loop.

7.6 Simulation and Experimental Verification

7.6.1 Simulation Result Analysis

In this section, the kinematic and dynamic models of the combination modality after the reconfiguration of the dual modular unit are established on the MATLAB R2021b numerical simulation platform, and the proposed control method for the combination modality and the control method to be compared are deployed on this basis for fault tolerance, disturbance rejection, and safety assurance multidimensional performance comparison verification. The total duration of the simulation is 120 s, with a step size of 1 ms. The kinematic reference trajectory is a straight line and a circular segmented trajectory, specifically, for $t < 80$, $v_{fd} = 0.4$ m/s, $w_{fd} = 0.0$ rad/s, for $t \geq 80$, $v_{fd} = 0.4$ m/s, $w_{fd} = 0.1$ rad/s, the actual reference trajectories x_{fd} and y_{fd} are updated according to $\dot{x}_{fd} = v_{fd}\cos(\theta_{fd})$, $\dot{y}_{fd} = v_{fd}\sin(\theta_{fd})$ and $\dot{\theta}_{fd} = w_{fd}$. To verify the adaptive fault tolerance capability of the combination modality, the experiment sets the health matrix of the combination modality member to have the following faults during operation

$$
\begin{aligned}
t < 40, \sigma_f &= \mathrm{diag}\{1, 1, 1, 1\},\\
t > 40, \sigma_f &= \mathrm{diag}\{1, 0, 0, 1\}.
\end{aligned}
$$

To verify the disturbance suppression capability in a noisy environment, the external disturbance and noise simulation of the velocity channel are set as

$$
\begin{aligned}
h_f &= \left[15\sin(t) + 3\cos(0.6t), 3\sin(t) + 0.5\cos(0.6t)\right]^{\mathrm{T}},\\
\xi_f &= \left[0.01\sin(1000t) + 0.005\cos(5000t), 0.01\cos(1000t) + 0.005\sin(5000t)\right]^{\mathrm{T}}.
\end{aligned}
$$

Finally, to verify the safety assurance capability of the proposed method, two obstacles are set on the motion path, and the positions and sizes of the obstacles are $(x_{1o}, y_{1o}, r_{1o}) = (5.0, 3.3, 1.0)$ m, $(x_{2o}, y_{2o}, r_{2o}) = (15.0, 7.5, 2.0)$ m.

This section designs three sets of comparative experiments in the simulation to verify the advantages of the proposed algorithm in fault tolerance, disturbance resistance, and safety assurance performance. All three sets of experiments are conducted under unknown disturbances. The first set of simulations verifies its fault tolerance capability through the fusion of self-adaptive fault tolerance and non-self-adaptive fault tolerance control schemes, respectively marked as C1 and C2; The second set of simulations uses the proposed AGFESO and traditional LESO control schemes introduced at the lower level to verify their observer compensation strategies' ability to handle dynamic level disturbances/noise/peaks, respectively marked as C3 and C4; The third set of simulations uses the self-adaptive safety barrier and Sharma's potential avoidance strategy at the kinematic level to demonstrate the effectiveness and smoothness advantages of the proposed safety strategy, respectively marked as C5 and C6. To ensure the fairness of the comparison, the same control gains are used at the kinematic and dynamic levels. The upper-level kinematic control gains are $k_{fx} = k_{fy} = 0.5$, $k_{f\theta} = 3$, $\kappa_{fx} = \kappa_{fy} = \kappa_{f\theta} = 4$, $\beta_{fx} = \beta_{fy} = \beta_{f\theta} = 8$, $\epsilon_{fx} = \epsilon_{fy} = 0.1$ and $\epsilon_{f\theta} = 0.5$, and the adaptive gains are $k_{fx} = k_{fy} = k^s_{fx} = k^s_{fy} = 100$, $k_{f\theta} = 50$, $\kappa^x_{fs} = \kappa^y_{fs} = 30$ and $\beta^x_{fs} = \beta^y_{fs} = 50$. The dynamic gain is $K_{f\eta} = \text{diag}\{2, 8\}$. The proposed AGFESO estimation gain parameters are $\omega_{1v} = \text{diag}\{10, 10\}$, $\omega_{2v} = \text{diag}\{490, 490\}$, $\tau_w = 0.1$ and $A_f = \text{diag}\{50, 50\}$, and the traditional LESO estimation gain is $L_1 = \text{diag}\{500, 500\}$. The self-adaptive safety barrier parameters are $\alpha^{ij}_{1fo} = 15$, $\alpha^{ij}_{2fo} = 30 (i = u, l; j = x, y)$, $\tau_{fo} = 0.05$, $\gamma_f = 0.4$, $\rho_f = 0.5$, $\epsilon_{fo} = 100$ and $c_{fo} = 100$. The gains related to Sharma's potential are optimized through repeated parameter tuning.

(1) Adaptive Fault Tolerance: Fig. 7.4 shows the comparison results of the proposed schemes with and without fault tolerance mechanisms under set fault conditions. As can be seen from the kinematic level trajectory tracking and error in Fig. 7.4a–b, without the fault tolerance mechanism, the trajectory tracking and precision significantly degrade due to the inappropriate distribution of wheel torques of the combination modality after the fault occurs. However, the control scheme that integrates adaptive fault tolerance can still maintain good trajectory tracking effects. Similarly, from the lower-level velocity control effects in Fig. 7.4c–f, it can be seen that the occurrence of faults will cause a complete degradation of velocity tracking, estimation, and its disturbance estimation. However, introducing adaptive fault tolerance continues to maintain the trajectory and velocity tracking control performance before the fault occurs. Figure 7.4f shows the torque response and fault status under two control situations. The actuation fault corresponds to a wheel torque output of 0, and the healthy wheel torque corresponds to an increase of about twice, indirectly showing the response status based on fault-tolerant triggering.

(2) Disturbance Estimation and Compensation: Fig. 7.5 clearly shows the comparison results of the proposed ESO and traditional ESO fusion schemes under the set fault conditions. As can be seen from Fig. 7.5e, in the presence of measurement noise and estimation initial errors, the traditional LESO easily triggers undesired transient peaks and sustained high-frequency oscillations, which will be directly reflected in the C4 torque output shown in Fig. 7.5f through the bottom feedforward compensation structure. After the fault occurs, the redistribution of torque will cause the healthy drive wheel to be responsible for outputting more torque, thereby reducing the ability and margin to compensate for disturbances, and the above two phenomena will directly destroy the velocity and trajectory tracking performance after the fault occurs, resulting in the overall performance degradation of C4 seen in Fig. 7.5a–d. In practical applications, the torque output generated relative to the peaking phenomenon and high-frequency oscillations shown in Fig. 7.5f can also easily cause the motor to lose control or even damage. The proposed AGFESO effectively reduces the initial transient peaks and noise high-frequency oscillations, and the high-quality disturbance estimation signal generated from it can better compensate for unknown lumped disturbances. The wheel torque relative to LESO also bears less, and in practical applications, a higher gain can be set relative to LESO to achieve better disturbance resistance.

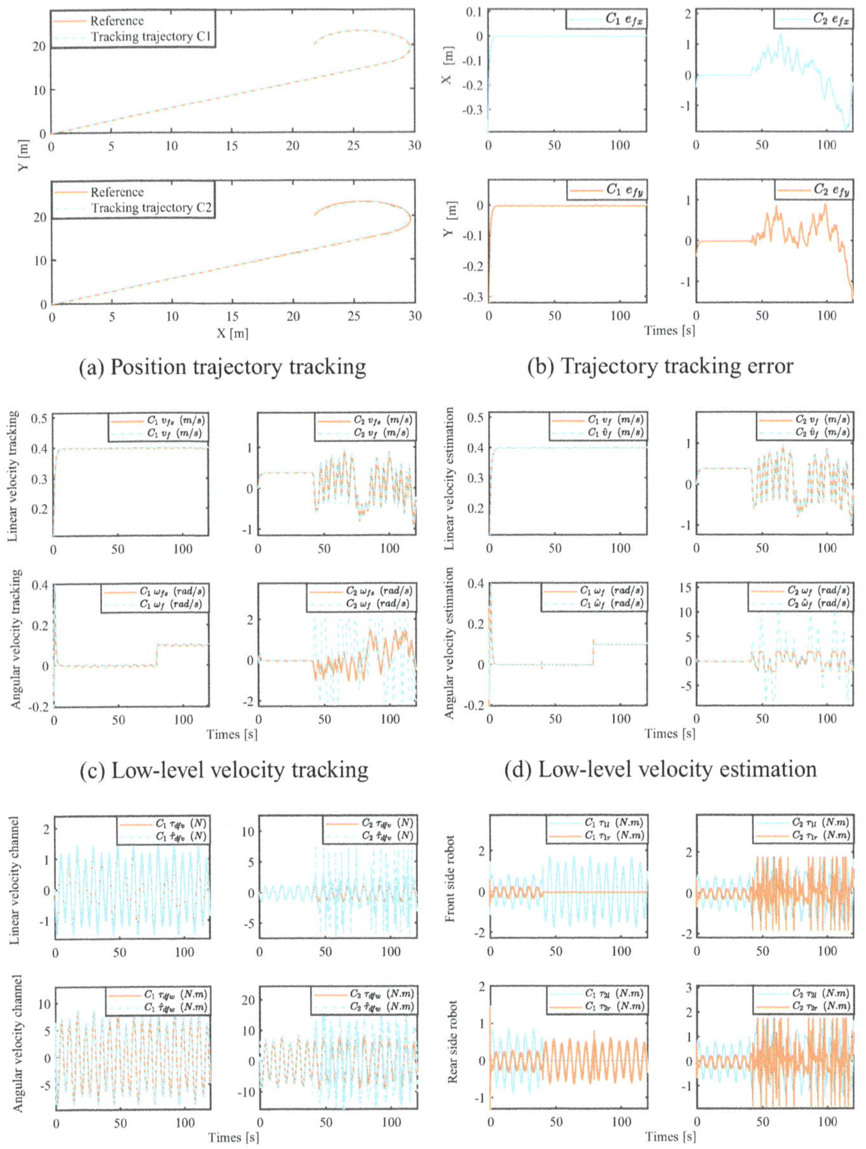

(a) Position trajectory tracking (b) Trajectory tracking error

(c) Low-level velocity tracking (d) Low-level velocity estimation

(e) Low-level disturbance estimation (f) Low-level torque output

Fig. 7.4 Fault-tolerant control effect simulation comparison of combination modality

(3) Obstacle Avoidance Safety Response: Fig. 7.6 compares the trajectory and velocity tracking effects of two safety evasion strategies in upper-level kinematics. As can be seen from the C5 and C6 position trajectory tracking and errors in

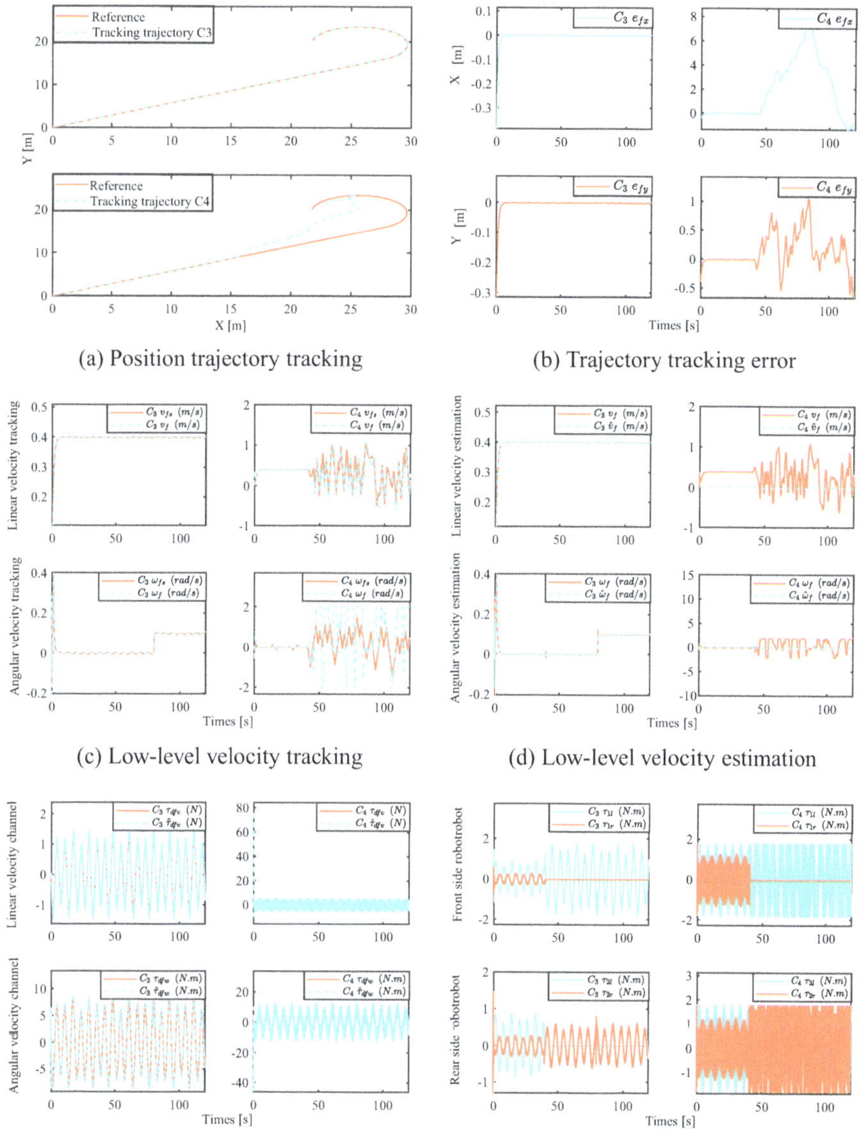

(a) Position trajectory tracking (b) Trajectory tracking error

(c) Low-level velocity tracking (d) Low-level velocity estimation

(e) Low-level disturbance estimation (f) Low-level torque output

Fig. 7.5 Disturbance rejection effect simulation comparison of combination modality

Fig. 7.6a–b, the adaptive safety barrier based on the mechanism of bypassing obstacles has a smaller oscillation amplitude during evasion, and a shorter recovery time after evasion, compared to the potential obstacle avoidance strategy of repelling obstacles. It has a clear advantage in smoothness, avoiding the motion stagnation

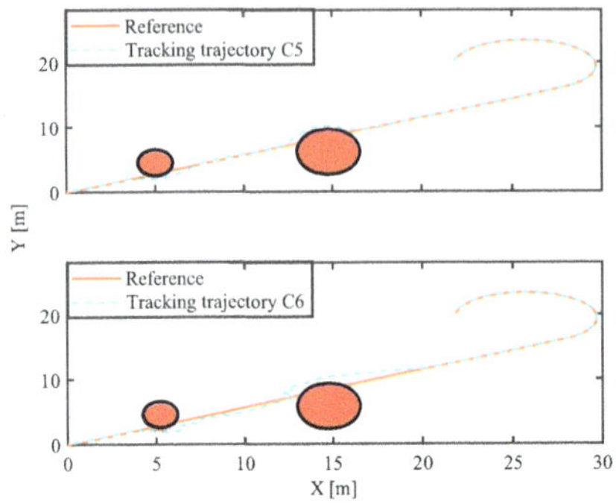

(a) Position trajectory tracking

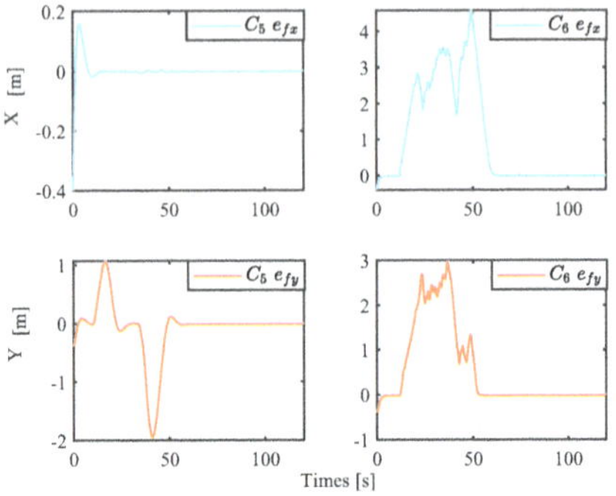

(b) Trajectory tracking error

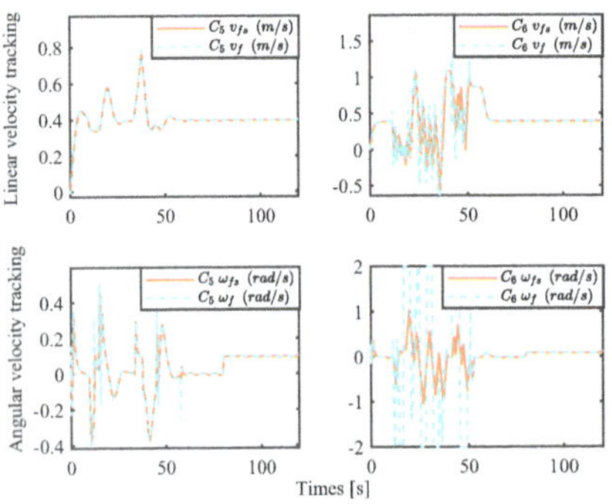

(c) Low-level velocity tracking

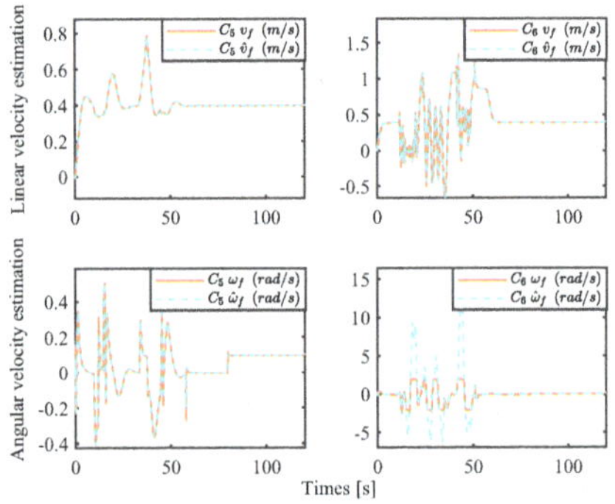

(d) Low-level velocity estimation

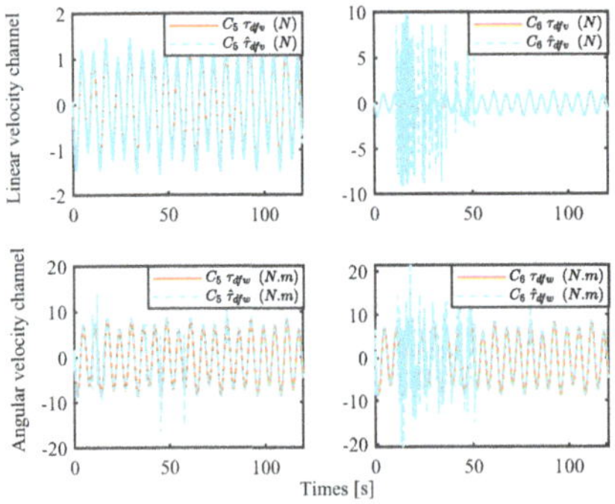

(e) Low-level disturbance estimation

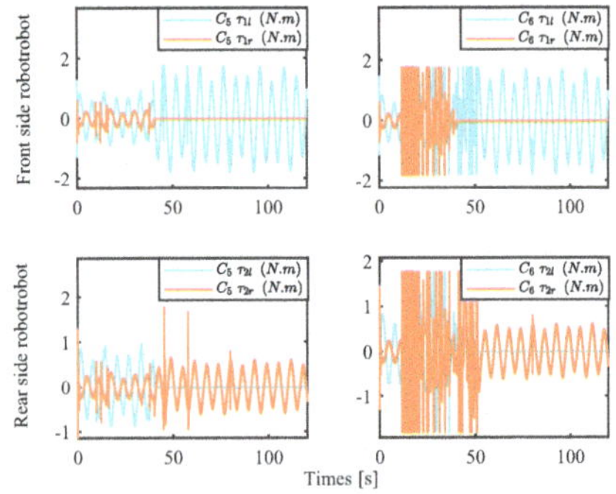

(f) Low-level torque output

Fig. 7.6 Comparison of safe control simulation effects of combination modality

and saturated oscillation that can easily occur with potential obstacle avoidance, and thus the C5 low-level response in Fig. 7.6c–f are also more moderate overall compared to C6. Therefore, the safety smoothness advantage of the adaptive safety barrier is more suitable for the motion control of the combined modality under the combined turning ability decline compared to the lumped potential evasion strategy.

7.6.2 Experimental Result Analysis

To verify the effectiveness of the proposed control of the combined modality, the combination modality control experiment verification platform is set up in Chap. 2, to carry out the verification of the fault tolerance, disturbance rejection, and safety adaptability advantages of the combination modality control. The experimental verification carrier is two-unit reconfiguration combination, which achieve velocity control at the dynamics level of the combination modality through the low-level motion control composed of actuation motors and microcontrollers, and use the motion capture camera array and image computer in the experimental site to recognize the markers attached to each unit and obstacles of the combination modality to calculate their positions and attitudes in the inertial system. The upper-level control computer obtains the status topic published by the image computer to construct the upper-level kinematic controller of the combination modality member in the reconfiguration swarm, and through the router, WiFi, and its own local communication network, it realizes the downlink of velocity commands and the reception of the low-level motion status of each unit of the combination modality.

To verify the fault tolerance, disturbance rejection, and safety assurance of the proposed combination modality control scheme, three sets of comparative experiments were co-designed with the simulation. The first set of experiments verifies the trajectory control performance of the proposed scheme with fault tolerance mechanism and fault removal mechanism by setting a complete failure of the forward unit and only providing power through the backward unit, marked as C1 and C2 respectively. The second set of experiments, based on the fault setting of the first set of experiments, introduces load disturbances respectively, and uses improved ESO and traditional LESO respectively in the bottom disturbance estimation compensation to verify the disturbance rejection of the combined modality control method under unknown disturbances, marked as C3 and C4 respectively. The last set of experiments pre-set obstacles on the motion trajectory of the combination modality, and the upper layer uses the proposed obstacle avoidance method and the potential-based Sharma obstacle avoidance method to compare and verify the smoothness of the safe evasion of the proposed method, marked as C5 and C6 respectively. The reference trajectory of the combination modality is a segmented trajectory of straight lines ($v_d = 0.06$ rad/s and $w_d = 0$ rad/s) and circles ($v_d = 0.06$ rad/s and $w_d = -0.03$ rad/s), with the saturation thresholds of the velocity commands being $\bar{v}_f = 1.5$, $\bar{w}_f = 1.5$, $\bar{w}_f = 1.5$, $\bar{a}_f = 0.5$ and $\bar{\omega}_f = 0.5$. To obtain the high-frequency state signals at the lower level, real-time linear/angular velocity feedback is obtained through wheel velocity conversion in the experiment. To ensure fairness in comparison, the upper-level kinematics and lower-level dynamics to be compared all use the same control gains. For the disturbance rejection comparison, due to the motor vibrations that may be induced in experimental applications, the AGFESO uses $\omega_{1v} = 10$ and $\omega_{2v} = 60$ as variable estimation gains, while LESO uses $\omega_{0v} = 10$ as a constant gain. For the safety comparison, the inner and outer triggering ranges of the artificial potential parameters optimized through repeated parameter tuning are $r_{in} = 0.15$ m and $r_{out} = 0.6$ m, the adjustment coefficients are $Q_{col} = 0.03$ and $\alpha_{col} = 0.05$, the safety barrier adjustment coefficients are $\alpha_{1fo}^{ij} = 20, \alpha_{2fo}^{ij} = 40 (i = u, l; j = x, y), \tau_{fo} = 0.1, \epsilon_{fo} = 100, c_{fo} = 0.1$, and the size and position of the obstacle are $(x_{1o}, y_{1o}, r_{1o}) = (0.85, 0.16, 0.15)$ m.

(1) Comparison of fault tolerance: Fig. 7.7 compares the trajectory tracking effects of the combination modality with and without adaptive fault tolerance. In Fig. 7.7a–b, the combination modality control method C1 with adaptive fault tolerance shows better transient performance and steady-state accuracy in reference trajectory tracking than the control method C2 without adaptive fault tolerance. Meanwhile, as seen in Fig. 7.7c, due to a more suitable torque distribution, which is beneficial for disturbance estimation and compensation, the velocity tracking effect

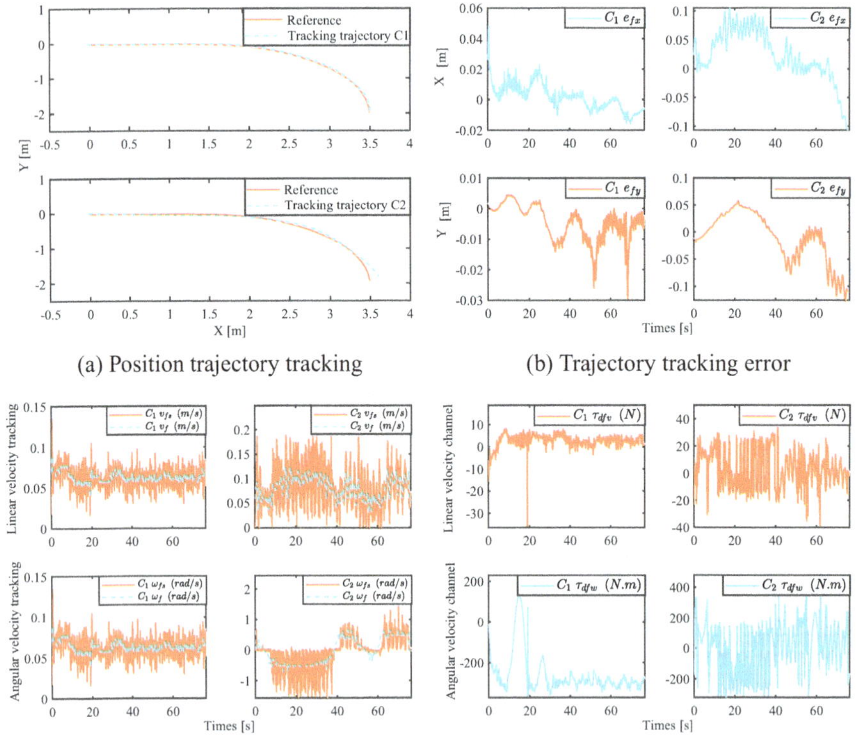

(a) Position trajectory tracking (b) Trajectory tracking error

(c) Low-level velocity tracking (d) Low-level disturbance estimation

Fig. 7.7 Comparison of fault-tolerant control experimental effects in combination modality of the control method with adaptive fault tolerance is also significantly better than that of the control method without a fault tolerance mechanism.

(2) Comparison of disturbance rejection performance: Fig. 7.8 compares the trajectory tracking effects of the combination modality integrating the proposed AGFESO and LESO respectively. Since the proposed AGFESO can set higher gains in the experiment to achieve better disturbance estimation effects, the control scheme C3 that integrates the ESO at the lower level will have better low-level velocity tracking effects as seen in Fig. 7.8c compared to the wheel velocity tracking in Fig. 7.8c. As can be seen from Fig. 7.8a–b, the control method C3 that integrates the proposed AGFESO is also significantly superior to the control method C4 that integrates LESO in terms of trajectory tracking and tracking error, and its tracking accuracy has not decreased compared to Fig. 7.7a–b. The disturbance estimation in C3 in Fig. 7.8d is also larger than the disturbance estimation in C1 in Fig. 7.7d due to the increase in load.

(3) Safety Evasion Performance Comparison: Fig. 7.9 compares the trajectory control effects of different evasion strategies at the kinematic level. Relative to the

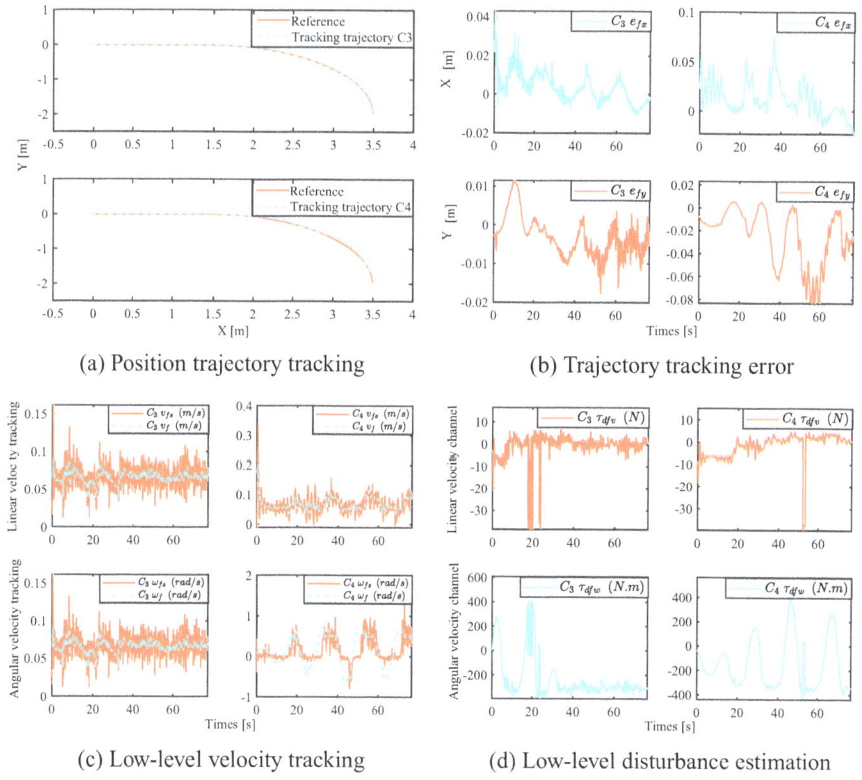

(a) Position trajectory tracking (b) Trajectory tracking error

(c) Low-level velocity tracking (d) Low-level disturbance estimation

Fig. 7.8 Comparison of robust control experimental effects in combination modality

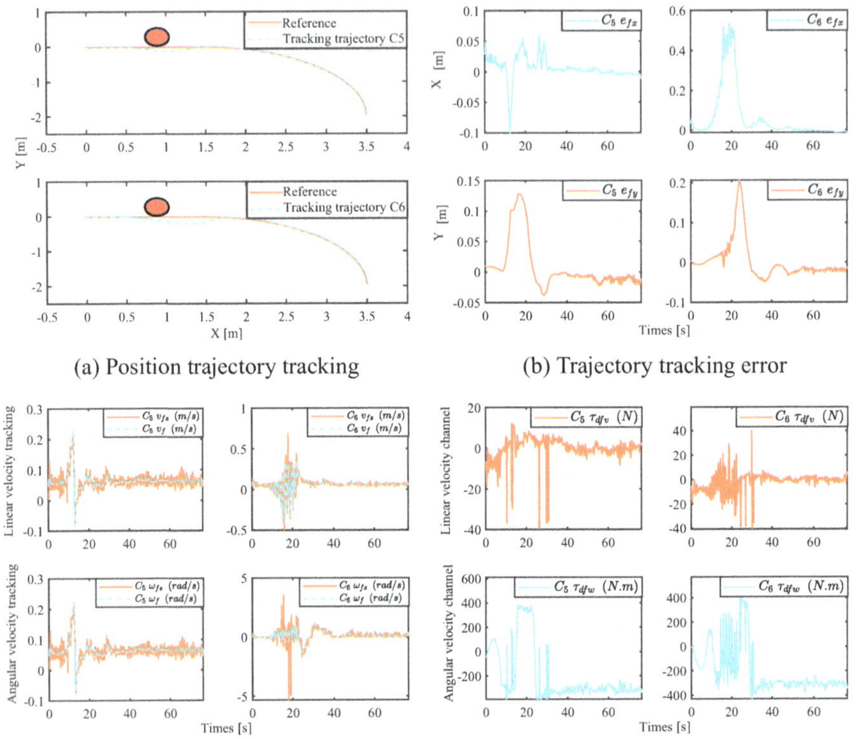

(a) Position trajectory tracking (b) Trajectory tracking error

(c) Low-level velocity tracking (d) Low-level disturbance estimation

Fig. 7.9 Safe control effect experimental comparison of combination modality

tracking trajectory based on potential evasion C6 in Fig. 7.9a, the tracking trajectory amplitude and recovery time based on the adaptive safety barrier C5 in Fig. 7.9a are clearly superior. As can be seen from Fig. 7.9b–c, C5 tracking error and velocity tracking also avoid the motion stagnation and violent oscillation phenomena that are prone to occur in potential obstacle avoidance C6 in Fig. 7.9b–c. Its superior safety and smooth characteristics are also more suitable for practical applications in controlling the movement of the combination modality in complex environments.

7.7 Summary

This article addresses the problem of basic control and adaptability enhancement of combination modality under the presence of actuation faults, unknown disturbances, and safety and saturation constraints. Firstly, this article constructs the kinematic and dynamic model of the combination modality under actuation faults and unknown uncertain disturbances, and adopts a control framework of the kinematic-dynamic cascade of the combination modality. A composite nonlinear feedback structure is used at the upper kinematic level to achieve basic position tracking, and a self-adaptive safety barrier is introduced to ensure more comprehensive safety of trajectory control. Secondly, a velocity controller based on the dynamics of the combination modality is designed at the lower dynamic level, and AGFESO and self-adaptive fault-tolerant mechanisms are embedded in the feedback control structure, respectively enhancing the robustness of velocity tracking under random actuation faults and lumped disturbances. Finally, the stability of kinematic and dynamic control is verified using Lyapunov's stability theorem, and negative factors such as faults, disturbances, and obstacles are gradually introduced in the simulation and experiment of the combination modality control, to comparatively verify the adaptability of the proposed control method.

References

1. Lashkari, N., Biglarbegian, M., Yang, S.X.: Development of a novel robust control method for formation of heterogeneous multiple mobile robots with autonomous docking capability. IEEE Trans. Autom. Sci. Eng. **17**(4), 1759–1776 (2020)
2. Shi, Y., Elara, M.R., Le, A.V., Prabakaran, V., Wood, K.L.: Path tracking control of self-reconfigurable robot hTetro with four differential drive units. IEEE Robot. Autom. Lett. **5**(3), 3998–4005 (2020)
3. Leggieri, S., Canali, C., Caldwell, D.G.: Design, modeling, and experimental analysis of the Crawler Unit for inspection in constrained space. Annu. Rev. Control. **57**, 100950 (2024)
4. Kayacan, E., Park, S., Ratti, C., Rus, D.: Learning-based nonlinear model predictive control of reconfigurable autonomous robotic boats: Roboats. In: Proceedings of 2019 IEEE/RSJ International Conference on Intelligence Robots System (IROS), pp. 8230–8237 (2019)
5. Gabrich, B., Li, G., Yim, M.: ModQuad-DoF: A novel yaw actuation for modular quadrotors. In: Proceedings of 2020 IEEE International Conference on Robotics Automation. (ICRA), pp. 8267–8273 (2020)
6. Ma, Y., Vincent, C., Najjar, M.E.B.E., Jiang, B.: Adaptive compensation of multiple actuator faults for two physically linked 2WD robots. IEEE Trans. Robot. **34**(1), 248–255 (2018)
7. Chwa, D.: Fuzzy adaptive tracking control of wheeled mobile robots with state-dependent kinematic and dynamic disturbances. IEEE Trans. Fuzzy Syst. **20**(3), 587–593 (2012)
8. Chwa, D.: Robust distance-based tracking control of wheeled mobile robots using vision sensors in the presence of kinematic disturbances. IEEE Trans. Ind. Electron. **63**(10), 6172–6183 (2016)
9. Chang, S., Wang, Y., Zuo, Z., Yang, H.: Fixed-time formation control for wheeled mobile robots with prescribed performance. IEEE Trans. Control Syst. Technol. **30**(2), 844–851 (2022)
10. Ma, Y., Cocquempot, V., Najjar, M.E.B.E.: Actuator failure compensation for two linked 2WD mobile robots based on multiple-model control. Int. J. Appl. Math. Comput. Sci. **27**(4), 763–776 (2017)

Chapter 8
Multimodal Behavior Integration and Scheduling Across Terrain Scenarios

Abstract Focusing on on-demand collaborative scheduling for reconfigurable swarms with multimodal capabilities, this work proposes a hierarchical decision-making architecture driven by both environmental and task layers. The architecture enables orderly transitions between different swarm modalities in response to obstacle traversal and fault events. A plug-and-play low-level controller library is developed to support the robust generation of task-sliced modal behaviours in complex scenarios. Two representative obstacle-crossing scenarios are presented to demonstrate the advantages of reconfigurable swarms in terms of on-demand reconfiguration and timely dispersion when facing various unexpected situations.

Keywords Finite state machine · Behaviour tree · Hierarchical scheduling · Plug-and-play · Multimodal control

8.1 Introduction

In Chaps. 4–7, we studied the coordination of reconfigurable swarm units, the combination of modular units for interface control, and the control of the combined modality. In this chapter, we design a framework for integrated scheduling of multimodal behavior for typical cross-terrain task scenarios of reconfigurable swarms, driven by scenario triggers. A large number of task planning strategies have been applied to the collaborative operations of robotic swarms [1–6], but research on task scheduling for reconfigurable swarms remains relatively limited. Moreover, existing scheduling frameworks tend to explore the adaptability of reconfigurable swarms in complex terrains and changing scenarios, making it difficult to expand the scale and structure of swarm reconfiguration [7] or lack detailed descriptions of bottom dynamic control [8], failing to fully exploit their advantages in scenario adaptability.

Focusing on the issue of smooth transition and scenario adaptability of multimodal behavior in reconfigurable swarms, this chapter combines the advantages of finite state machines and behavior tree decision-making methods to design a hierarchical structure of multimodal behavior scheduling framework. This enables spontaneous smooth transitions of complex modalities in reconfigurable swarms. Based

T. Jiang et al., *Multi-modality Control of Reconfigurable Swarm*,
https://doi.org/10.1007/978-981-95-0305-6_8

on the model characteristics of each modality in Chap. 2, we deeply integrate the results of modality internal control and modality switching studies from Chaps. 3–7. We improve the coordination of modular units, promote the docking of modular units/combined modalities, the progress of combined modalities, the coordination of modular units and combined modalities, and the coordination of combined modalities. We specify the control forms of behavior tree action nodes, pushing the task coordination and localization application of reconfigurable swarms under any reconfiguration scale across terrain scenarios. Finally, this chapter verifies the effectiveness of the overall motion control framework of multimodal reconfigurable swarms through the construction of two specific cross-terrain task scenarios, highlighting its advantages in balancing efficiency and adaptability to complex scenarios.

8.2 Reconfigurable Swarm Multimodal Behavior Scheduling Framework

8.2.1 Overall Approach to Multimodal Behavior Model Construction and Scheduling Framework Design

Reconfigurable swarms can achieve capability enhancement and fault recovery through the recombination of different modular units/combination modalities, making them more adaptable to cooperative tasks in complex scenarios compared to traditional swarms with fixed/weakly variable configurations. This chapter takes two typical cross-terrain scenarios as examples, integrates swarming, docking, and combination control methods on the basis of existing hardware, and explores on-the-spot adjustment of multimodal behavior and smooth transition of modalities. Finite state machines and behavior trees are two mainstream decision-making methods for swarm behavior scheduling. A finite state machine is a mathematical model describing a finite number of state systems and their state transition rules, usually composed of initial states, states, events, transition functions, and termination states, and is very suitable for describing the states of swarm members and entities and their transition rules; however, its interdependence between states is severe, the coupling degree is too high, and the structural extensibility is poor. Behavior trees represent complex behavior logic through nodes and sub-nodes, have low coupling, easily control conflicts between nodes, and are convenient for constructing refined behavior models; although behavior trees can describe behavior sequences and transition rules at the same time, their logical structure is too complex, and the design of behavior models is difficult. Therefore, the multimodal behavior model of the reconfigurable swarm in this section adopts the hierarchical behavior control framework of finite state machines and behavior trees as shown in Fig. 8.1, describing the tasks and their transition rules of the reconfigurable swarm across terrain scenarios according to modality division through finite state machines, and the behavior subtrees under each task state focus more on representing the specific behavior logic of different tasks.

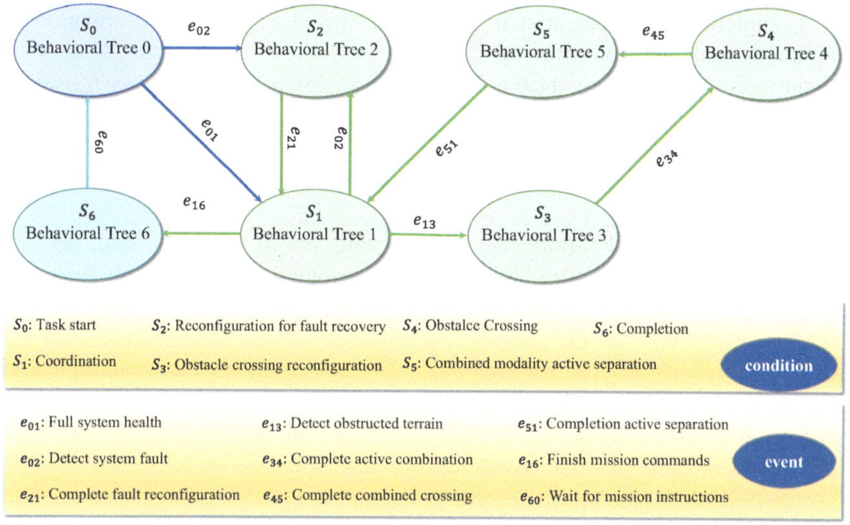

Fig. 8.1 Hierarchical behavior scheduling framework

The above framework design not only retains the advantages of finite state machines and behavior trees in behavior scheduling, but also avoids their shortcomings in single application in complex task coordination. Reconfigurable swarms will encounter varying terrains and sudden situations in complex scene coordination, such as: (1) Randomly appearing impassable dangerous terrains, hindering the advancement of the unit swarm; (2) Various malfunctions may occur in the unit swarm, resulting in loss of mobility, etc. When responding to specific tasks, the reconfigurable swarm will adaptively switch to the task state and specific modality suitable for the current scene according to the set trigger events.

The behavior framework designed in this section has 7 task states as shown in Fig. 7.1, including:

① Task start (S_0): The unit swarm is in the task preparation stage;

② Coordination (S_1): The unit or combined swarm performs coordinated tasks according to the planned tasks;

③ Reconfiguration for fault recovery (S_2): The faulty unit in the swarm and the nearest healthy unit or combined modality in the adjacent network are reconfigured into a combined modality with mobility;

④ Obstacle crossing reconfiguration (S_3): Healthy units or combined modalities are orderly reconfigured into suitable combined modality modalities to form the current scene-adaptive capability;

⑤ Obstacle crossing (S_4): Use the combined modality that adapts to the scene to cross the current obstacle terrain with combined movement;

⑥ Combined modality active separation (S_5): The combined modality separates into healthy units or smaller scale combined modalities before the combination;

⑦ Completion (S_6): The unit or combined swarm reaches the destination to complete the task and stops moving.

In the finite state machine, there are 9 events that can trigger the adaptive transition of various task states in the reconfigurable swarm across terrains when encountering changes in the external environment and its own modality. They are as follows:

① System is fully healthy (e_{01}): All units are in a healthy state, triggering the transition from S_0 to S_1;

② System has partial failure (e_{02}): Some units are in a failure state, but can still recover their coordination ability through reconfiguration for fault recovery, triggering the transition from the S_0 or S_1 to S_2;

③ Reconfiguration for fault recovery completed (e_{21}): The failed unit and the healthy units in the adjacent network successfully reconfigure into a combination that can recover its ability, triggering the transition from S_2 to S_1;

④ Encounter obstacle terrain (e_{13}): The unit swarm encounters an insurmountable obstacle terrain, triggering the transition from S_1 to S_3;

⑤ Obstacle crossing reconfiguration completed (e_{34}): The unit or combination completes the obstacle crossing reconfiguration docking, triggering the transition from S_3 to S_4;

⑥ Depart from obstacle terrain (e_{45}): The combination crosses the obstacle terrain, triggering the transition from S_4 to S_5;

⑦ Active separation completed (e_{51}): Each unit or small-scale combination in the combination separates to the swarm formation position, triggering the transition from S_1 to S_5, ending the movement.

⑧ Arrival at the destination area (e_{16}): Each unit reaches the target area, triggering the transition of S_1 to S_6, and terminating the movement;

⑨ Receiving task instructions (e_{60}): The reconfigurable swarm receives new coordination tasks, triggering the transition of S_6 to S_0.

8.2.2 *Design of Finite State Machine Behavior Subtree in the Coordination Behavior Framework*

(1) Design of Behavior Subtree Nodes

In the hierarchical framework, each state of the finite state machine has a behavior subtree, including sequence, selection, parallel, condition, action nodes, etc., which promote behavior actions in a planned and step-by-step manner. The behavior trees of task initiation and completion states are shown in Fig. 8.2, and the behavior trees of different task states such as coordination, fault recovery under combined modality, obstacle crossing reconfiguration, obstacle crossing, separation, etc., are shown in Fig. 8.3.

The task initiation state includes a behavior tree of self-check, which determines through the selection node whether the system is fully healthy, partially faulty, and

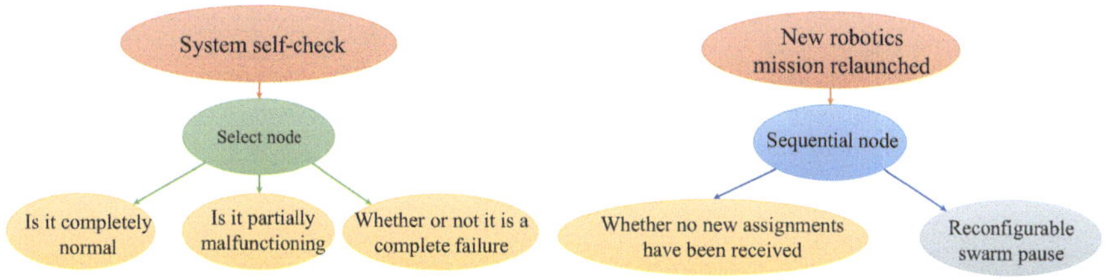

(a) Subbehavior tree of RS task start (b) Subbehavior tree of Rs completion

Fig. 8.2 Subbehavior tree of RS task initiation and completion

completely faulty, locates the current health status of the swarm system. The mechanism ensures that the swarm state transitions to the coordination or fault recovery under combined modality state in the finite state machine, or terminates the coordination task under complete failure.

The reconfigurable coordination behavior tree executes system self-check (selection node), coordination formation (action node), and whether to reach the endpoint (condition node) simultaneously through the parallel node. When the endpoint is not reached or the health status changes, it always executes the current coordination formation action.

The reconfigurable swarm fault recovery under combined modality behavior tree sequentially executes judgments through sequential nodes, determining whether there are healthy unit or combination modalities in the adjacent network of the faulty unit (conditional node), selecting the nearest healthy unit or combination for active reconfiguration (action node), and judging whether the fault recovery under combined modality is successful (conditional node). After reconfiguring with the nearby healthy unit/combination, the behavior capability is restored, and it is determined that the fault recovery under combined modality is successful.

The reconfigurable swarm obstacle crossing reconfiguration behavior tree sequentially executes judgments through sequential nodes, determining whether there are nearby healthy unit or combination modalities in the adjacent network of the unit to be interfaced (selection node), judging whether the obstacle crossing reconfiguration has been successful (selection node), and multiple units actively and orderly reconfigure (action node). In the case of having adjacent healthy unit or combination modalities and no obstacle crossing reconfiguration, the reconfiguration among multiple units is continuously executed.

The reconfigurable swarm combined obstacle behavior tree sequentially executes judgments through sequential nodes, determining whether it has not deviated from the obstacle formation (selection node), and the combined modality overcomes the obstacle in coordination or advances (action node). In the case of not deviating from the obstacle terrain, the action of obstacle crossing is always executed.

The reconfigurable swarm active separation behavior tree sequentially executes judgments through sequential nodes, determining whether it has not separated into

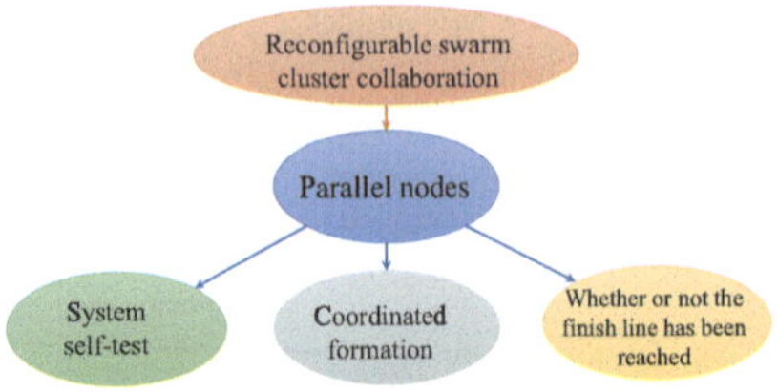

(a) Subbehavior tree of RS cooperation

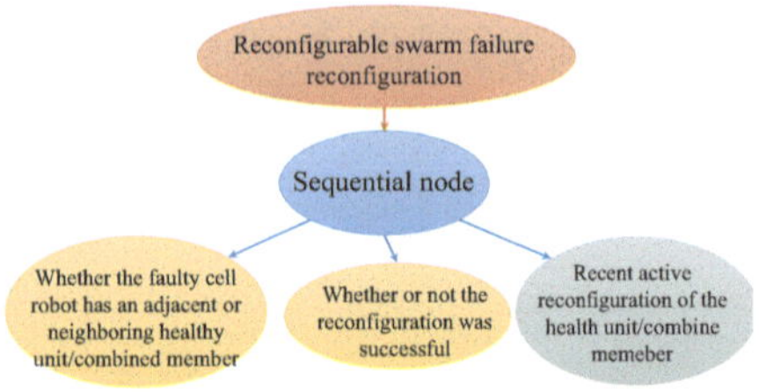

(b) Subbehavior tree of RS for fault recovery

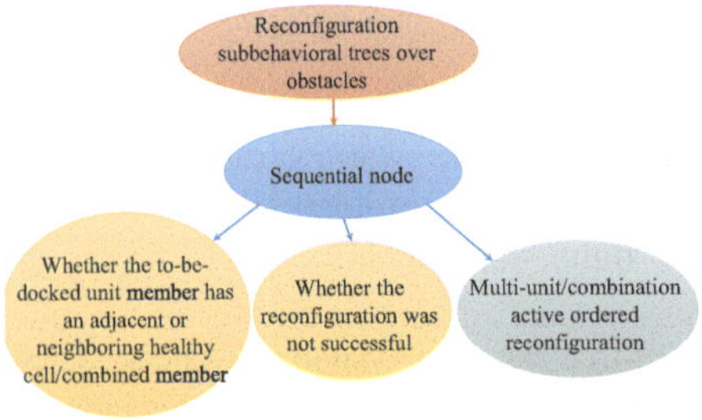

(c) Subbehavior tree of RS for fault recovery

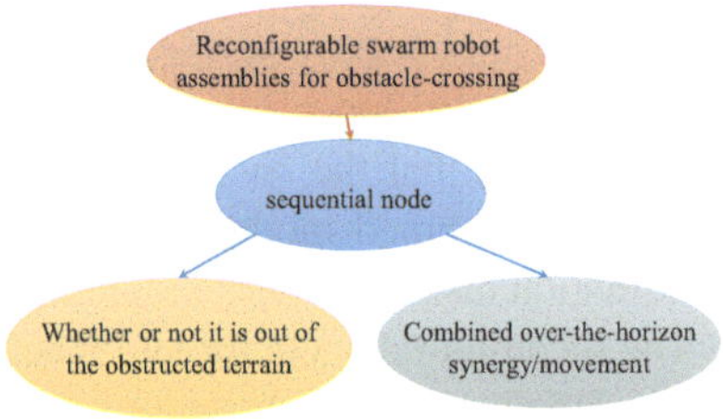

(d) Subbehavior tree of RS obstacle crossing

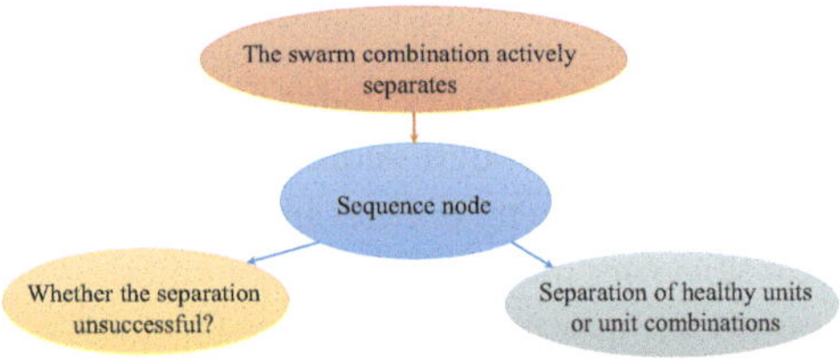

(e) Subbehavior tree of RS active separation

Fig. 8.3 Subbehavior tree of multi-task collaboration in RS

a formation (selection node), and the combined modality separates to the formation position (action node). When each unit or combination has not reached the predetermined separation position, the action of separation movement is always executed.

The task completion sub-behavior tree sequentially executes judgments through sequence nodes to determine whether a new task has been received (selection node), and whether the swarm action is paused (action node). When each unit does not receive a new task, it always remains in a paused state.

(2) Description of the Specific Control Form of the Action Node

The coordinated actions of the entire swarm may involve specific control forms such as monomer swarms, monomer/combined modality reconfiguration docking, monomer and combined modality swarms, combined modality and combined modality swarms, and combined modality movement. This section combines the results of modal control and switching studies from Chaps. 4–7, encapsulates the above swarm reconfiguration and modal control methods, and constructs a multi-modal motion controller set that can be polled and called by the behavior subtree action nodes of the scheduling framework shown in Fig. 8.4, coordinating with scheduling rules to complete complex scene adaptation and swarm task coordination. For example, the monomer swarm adopts kinematic and dynamic hierarchical controllers respectively using Eqs. (5.25) and (5.26) from Chap. 5 and Eqs. (4.25) and (4.41) from Chap. 4; the upper-level kinematic models of the monomer and combined modality are basically consistent, to simplify the form of the controller set, the active docking monomer can adopt the controller form of the monomer for the combined modality and the monomer in the swarm, but the lower-level dynamic forms are not compatible, requiring the use of targeted controller forms, therefore, the kinematic level docking of the monomer and combined modality adopts Eqs. (6.85) and (6.99) from Chap. 6 in all stages, and the lower-level dynamics adopt Eqs. (6.25), (6.26), (6.31) and (6.65) from Chap. 6 and Eqs. (7.44), (7.48) and (7.57) from Chap. 7; the kinematics of the monomer and combined modality swarm movement adopt Eqs. (5.25) and (5.26) from Chap. 5, and the dynamics of the monomer and combined modality adopt Eqs. (4.25) and (4.41) from Chap. 4 and Eqs. (7.44), (7.48) and (7.57) from Chap. 7; the kinematics of the combined modality and combined modality swarm adopt Eqs. (5.25) and (5.26) from Chap. 5, and the dynamics adopt Eqs. (7.44), (7.48) and (7.57) from Chap. 7; the kinematics and dynamics of the combined modality movement adopt Eqs. (7.39)–(7.41) and Eq. (7.44), (7.48) and (7.57) from Chap. 7 respectively. At the same time, because the differences in the dynamic descriptions of the unit modality and the combined modality modal control model are mainly at the lower level, and the kinematic descriptions at the upper level are basically consistent, a consistent kinematic controller form can also be adopted at the upper level of the unit modality and the combined modality modal. According to the reconfigurable swarm, the corresponding dynamic model is adopted according to the reconfiguration scale. The control tasks of the reconfigurable swarm can be mainly divided into two categories: trajectory tracking of each modality and multi-modal formation. The following integrates the control strategies from Chap. 4–7, and gives detailed controller forms for the unit modality and combined modality of the reconfigurable swarm.

First is the kinematic level, where the unit and combination modalities can adopt the following identical controller forms.

① Trajectory Control Under Docking and Obstacle Crossing Maneuvers

Similar to the previous sections, the desired trajectory during modality switching and after transformation is set to x_d, y_d, and the adaptive safety barrier is used to solve the safety avoidance requirement. Similar to Chaps. 4 and 7, a position-loop and attitude-loop cascaded upper-level kinematic controller form is adopted. v_{ix} and v_{iy} are the longitudinal and lateral velocity control commands, the subscript i can be $1 \cdots N$ or f, representing the swarm unit number or combined modality, v_{id} is the linear velocity control command, and the corresponding relationship between the linear velocity control command and the lateral/longitudinal velocity control command is as follows

$$v_{id} = \cos\theta_i v_{is}^x + \sin\theta_i v_{is}^y = \cos(\theta_i - \theta_{id}) \sqrt{\left(v_{is}^x\right)^2 + \left(v_{is}^y\right)^2}, \tag{8.1}$$

where, θ_{id} is the desired heading angle, which can be derived from $\theta_{id} = \arctan 2\left(v_{is}^y, v_{is}^x\right)$.

The design of the unit modality and combined modality trajectory control kinematic level universal position-loop longitudinal and lateral control laws are

$$v_{ix} = -k_{ix} \tanh\left(\frac{x_{ie}}{k_{ix}}\right) - k_{is}^x \lambda_{is}^x + \dot{x}_d, \tag{8.2}$$

$$v_{iy} = -k_{iy} \tanh\left(\frac{y_{ie}}{k_{iy}}\right) - k_{is}^y \lambda_{is}^y + \dot{y}_d, \tag{8.3}$$

Here, $k_{ix}, k_{iy} \in \mathbf{R}_+$ are the controller design gains, $x_{ie} = x_i - x_d - \gamma_i \tanh \lambda_{ix}$ and $y_{ie} = y_i - y_d - \gamma_i \tanh \lambda_{iy}$ are the longitudinal and lateral control errors, $\lambda_{is}^x, \lambda_{is}^y$ are auxiliary dynamic variables, similar to the safety protection mechanism in Chap. 5, used to improve the adaptive auxiliary dynamic system caused by the oscillation of the safety fence

$$\begin{aligned} \dot{\lambda}_{is}^x &= \frac{\cosh^2\left(\lambda_{is}^x\right)\left(-k_{is}^x \lambda_{is}^x\right)\lambda_{is}^x + \Delta_{is}^x}{\gamma_i}, \\ \dot{\lambda}_{is}^y &= \frac{\cosh^2\left(\lambda_{is}^y\right)\left(-k_{is}^y \lambda_{is}^y\right)\lambda_{is}^y + \Delta_{is}^y}{\gamma_i}, \end{aligned} \tag{8.4}$$

where $k_{is}^x, k_{is}^y, \gamma_i \in \mathbf{R}_+$ are adaptive gains.

The attitude error is defined as $\theta_{ie} = \theta_i - \theta_{id} - \gamma_{i\theta} \tanh \lambda_{i\theta}$, and the attitude loop control law designed to stabilize the attitude tracking error is

$$w_{id} = -k_{i\theta}\theta_{ie} + \dot{\theta}_{id} - k_{is}^\theta \lambda_{i\theta}, \tag{8.5}$$

where $k_{i\theta} \in \mathbf{R}_+$ is the controller design gain, $\lambda_{i\theta}$ is an auxiliary variable, generated by the adaptive auxiliary dynamic signal that suppresses the saturation of the angular velocity command

$$\dot{\lambda}_{i\theta} = \cosh^2\left(\lambda_{i\theta}\right)\left(-k_{is}^\theta \lambda_{i\theta} + \Delta_{iw}\right)/\gamma_{i\theta}, \tag{8.6}$$

where $k_{is}^\theta, \gamma_{i\theta} \in \mathbf{R}_+$ are adaptive gains.

② swarm formation control under hybrid coordination modality

In the coordination, the reconfigurable swarm hybrid modalities with modular unit and combined structure of different reconfiguration scales, the unit modality and the combined modality have similar kinematic models, so the controller form is consistent with the formation controller form in Chap. 4, the difference is that the formation members are unit modality or combined modality of different reconfiguration scales.

The virtual longitudinal and lateral velocities v_{0x} and v_{0y} (longitudinal and lateral velocity commands) in the kinematics of the leader are designed as

$$\begin{aligned} v_{0x} &= -k_{0x}\tanh\left(\frac{e_{0x}}{k_{0x}}\right) + \dot{x}_r - k_{0s}^{x}\lambda_{0x}, \\ v_{0y} &= -k_{0y}\tanh\left(\frac{e_{0y}}{k_{0y}}\right) + \dot{y}_r - k_{0s}^{y}\lambda_{0y}, \end{aligned} \tag{8.7}$$

Here, k_{0x} and k_{0y} are the control gains of the leader, x_r, y_r are the desired trajectories of the leader, $e_{0x} = x_0 - x_r - \gamma_0 \tanh(\lambda_{0x})$ and $e_{0y} = y_0 - y_r - \gamma_0 \tanh(\lambda_{0y})$ are the control errors, λ_{0x}, λ_{0y} are the auxiliary signals, generated by the adaptive auxiliary dynamic system that suppresses command saturation

$$\begin{aligned} \dot{\lambda}_{0x} &= \cosh^2(\lambda_{0x})\left(-k_{0s}^{x}\lambda_{0x} - \Delta_{0v}\cos(\theta_0)\right)/\gamma_0, \\ \dot{\lambda}_{0y} &= \cosh^2(\lambda_{0y})\left(-k_{0s}^{y}\lambda_{0y} - \Delta_{0v}\sin(\theta_0)\right)/\gamma_0, \end{aligned} \tag{8.8}$$

where $k_{0s}^{x}, k_{0s}^{y}, \gamma_0 \in \mathbf{R}_+$ are the adaptive gains.

The guidance direction (desired direction) and virtual linear velocity θ_{0d} and v_{0d} (linear velocity command) can be expressed as

$$\begin{aligned} \theta_{0d} &= \operatorname{atan2}\left(v_{0y}, v_{0x}\right), \\ v_{0d} &= \cos(\theta_0 - \theta_{0d})\sqrt{v_{0x}^2 + v_{0y}^2}, \end{aligned} \tag{8.9}$$

where $\lambda_{0\theta}$ is the auxiliary variable of the leader's direction; the corresponding adaptive auxiliary dynamic system is shown below

$$\dot{\lambda}_{0\theta} = \cosh^2(\lambda_{0\theta})\left(-k_{0\theta}\lambda_{0\theta} - \Delta_{0w}\right)/\gamma_{0\theta}, \tag{8.10}$$

where $\Delta_{0w} = w_{0d} - w_{0s}$; $\gamma_{0\theta}$ and $k_{0\theta}$ are the adaptive gains.

In order to track the leader's virtual direction, the virtual angular velocity w_0 (angular velocity command) is designed as

$$w_{0d} = -k_{0\theta}\theta_{0e} + \dot{\theta}_{0d} - k_{0s}^{\theta}\lambda_{0\theta}, \tag{8.11}$$

where $\theta_{0e} = \theta_0 - \theta_{0d} - \gamma_{0\theta}\tanh(\lambda_{0\theta})$, $k_{0\theta} \in \mathbf{R}_+$ is the control gain.

The longitudinal and lateral velocity commands v_{ix} and v_{iy} of the ith formation follower ($i = 1, \ldots, m$) are designed as follows

$$\begin{aligned} v_{ix} &= -k_{ix}\tanh\left(\frac{e_{ix}}{k_{ix}}\right) + \hat{h}_{ix} + \dot{p}_{ix} - k_{is}^{x}\lambda_{ix}, \\ v_{iy} &= -k_{iy}\tanh\left(\frac{e_{iy}}{k_{iy}}\right) + \hat{h}_{iy} + \dot{p}_{iy} - k_{is}^{y}\lambda_{iy}, \end{aligned} \tag{8.12}$$

where

$$\dot{\hat{h}}_{ix} = -k_{ih}^{x}\left(\sum_{j=1}^{m} a_{ij}\left(\hat{h}_{ix} - \hat{h}_{jx}\right) + b_i\left(\hat{h}_{ix} - \dot{x}_0\right)\right),$$

$$\dot{\hat{h}}_{iy} = -k_{ih}^{y}\left(\sum_{j=1}^{m} a_{ij}\left(\hat{h}_{iy} - \hat{h}_{jy}\right) + b_i\left(\hat{h}_{iy} - \dot{y}_0\right)\right),$$

Here, k_{ix} and k_{iy} are the formation control gains of the ith follower, $\hat{h}_{ix}$ and $\hat{h}_{iy}$ are the estimated values of the leader's longitudinal and lateral velocity, k_{ih}^{x} and k_{ih}^{x} are the observation gains, $z_{ix} = x_i - p_{ix} - x_0 - \gamma_i \tanh(\lambda_{ix})$ and $z_{iy} = y_i - p_{iy} - y_0 - \gamma_i \tanh\left(\lambda_{iy}\right)$ are the tracking errors of the follower, $e_{ix} = \sum_{j=1}^{m} a_{ij}\left(z_{ix} - z_{jx}\right) + b_i z_{ix}$ and $e_{iy} = \sum_{j=1}^{m} a_{ij}\left(z_{iy} - z_{jy}\right) + b_i z_{iy}$ are the formation coordination errors, $\lambda_{ix}, \lambda_{iy}$ are the auxiliary signals, generated by the adaptive auxiliary dynamic system that suppresses command saturation

$$\begin{aligned} \dot{\lambda}_{ix} &= \cosh^2(\lambda_{ix})\left(-k_{is}^{x}\lambda_{ix} - \Delta_{iv}\cos(\theta_i)\right)/\gamma_i, \\ \dot{\lambda}_{iy} &= \cosh^2\left(\lambda_{iy}\right)\left(-k_{is}^{y}\lambda_{iy} - \Delta_{iv}\sin(\theta_i)\right)/\gamma_i, \end{aligned} \tag{8.13}$$

where $\Delta_{iv} = v_{id} - v_{is}$; $k_{is}^{x}, k_{is}^{y}, \gamma_i$ are the adaptive gains of the ith follower; v_{is} is the saturation virtual linear velocity of the ith follower under nested saturation.

The virtual heading (reference heading) and virtual linear velocity of the ith $(i = 1, \ldots, m)$ follower can be written as

$$\begin{aligned} \theta_{id} &= \operatorname{atan2}\left(v_{iy}, v_{ix}\right), \\ v_{id} &= \cos(\theta_i - \alpha_i)\sqrt{v_{ix}^2 + v_{iy}^2}, \end{aligned} \tag{8.14}$$

where $\lambda_{i\theta}$ is the heading auxiliary variable of the ith follower; the corresponding adaptive auxiliary dynamic system is shown as follows

$$\dot{\lambda}_{i\theta} = \cosh^2(\lambda_{i\theta})\left(-k_{is}^{\theta}\lambda_{i\theta} - \Delta_{iw}\right)/\gamma_{i\theta}, \tag{8.15}$$

where $\Delta_{iw} = w_{id} - w_{is}$; $\gamma_{i\theta}$ and $k_{is}^{\theta} \in \mathbf{R}_{+}$ are the adaptive gains.

In order to track the virtual heading of the ith follower, the virtual angular velocity is designed as follows

$$w_{id} = -k_{i\theta}\tanh\left(\frac{\theta_{ie}}{k_{i\theta}}\right) + \dot{\theta}_{id} - k_{i\theta}\lambda_{is}^{\theta}, \tag{8.16}$$

where $\theta_{ie} = \theta_i - \theta_{id} - \gamma_{i\theta}\tanh(\lambda_{i\theta})$; $k_{i\theta} \in \mathbf{R}_{+}$ represents the control gain. At the dynamics level, both the modular unit modality and the combined modality aim to track the velocity command from the upper level. Both can adopt the state feedback and disturbance feedforward compensation structure based on their respective model forms (3.15), (7.12). The disturbance estimation can adopt the adaptive gain filter extended state observer (7.44) in Chap. 7. The basic form of the controller for the modular unit modality and the combined modality is as follows.

For the modular unit modality, each member of the formation integrates the dynamics model and disturbance estimation feedforward compensation. The lower-level dynamics controller is designed as

$$u_i = \left(\bar{M}_{1i}^{-1}\bar{B}_{1i}\right)^{+}\left(-K_{i\eta}\eta_{ie} + \dot{\eta}_{id} - F_i(\cdot) - \hat{x}_{2i}\right), \tag{8.17}$$

where $K_{i\eta} = \text{diag}\{k_{iv}, k_{iw}\} \in \mathbf{R}_{+}^{2}$ is the diagonal gain matrix, $\eta_{ie} = \eta_i - \eta_{id}$ is the velocity tracking error. In this case, the combined modality under the cascaded structure is subject to nested saturation dynamics control input $\eta_{id} = [v_{is}, w_{is}]^{\mathrm{T}}$ is the desired signal of the lower-level velocity dynamics, and the control input u_i can be applied through torque distribution. The modular unit modality does not have redundant drives. When a fault occurs, it is necessary for the healthy modular unit modality to reconfigure and restore its action capability.

For the modular unit modality in the swarm, for each $\sigma_{f(k)}$ fault situation, the combined modality dynamics controller integrating the dynamics model and disturbance estimation feedforward compensation is designed as

$$u_{f(k)} = \left(\bar{M}_{1f}^{-1}\bar{B}_{1f}\sigma_{f(k)}\right)^{+}\left(-K_{f\eta}\eta_{fe} + \dot{\eta}_{fd} - F_f(\cdot) - \hat{x}_{2f}\right), \tag{8.18}$$

where $K_{f\eta} = \text{diag}\{k_{fv}, k_{fw}\} \in \mathbf{R}_{+}^{2}$ is the diagonal gain matrix, $\eta_{fe} = \eta_f - \eta_{fd}$ is the velocity tracking error. In this case, the combined modality under the cascaded structure is subject to nested saturation dynamics control input $\eta_{fd} = \left[v_{fs}, w_{fs}\right]^{\mathrm{T}}$ is the desired signal of the lower-level velocity dynamics.

8.3 Cross-Terrain Scenario Construction and System Verification of Rcconfigurable Swarm

In this section, through two typical cross-terrain scenarios, namely the collaboration across floors by a three-unit reconfigurable swarm and the outdoor terrain collaboration by a four-unit reconfigurable swarm, the smooth transition of multi-modalities in task collaboration, the balance of the swarm's capability aggregation and dispersion as needed, and the characteristic advantages of "throughput capability" and "execution efficiency" in application scenarios are fully verified.

(1) Verification of the Scenario of Collaboration Across Floors by a Three-Unit Reconfigurable Swarm

The construction and scheduling process of cross-floor collaborative scenarios is shown in Fig. 8.3. The overall behavior process is autonomously triggered through

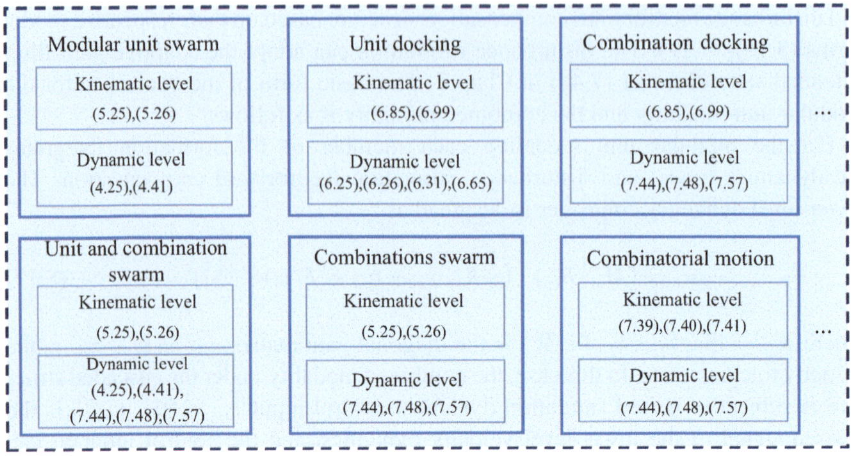

Fig. 8.4 RS modality motion controller set

the aforementioned modal behavior scheduling framework. The positions and poses of each unit and obstacles in the field can be identified by attaching markers and using image computing to calculate. Each unit is equipped with a WiFi module and the master control computer uses a router to establish a local communication network for information sharing. Start points, destination points, and building obstacle terrain markers are placed at different locations in the field, and square boxes are placed as preset environmental obstacles to simulate floor terrain elements and trigger the modal transition and action node safety response of the reconfigurable swarm. Finally, by integrating all the above settings, the master control computer processes the multi-modal behavior integrated scheduling and action node control code for cross-floor collaboration, verifying the advantages of the reconfigurable swarm's situational response, multi-modal coordination, and scenario adaptation.

In the preset scenario of cross-floor collaboration of the reconfigurable swarm, the swarm size is set to three units. The overall coordination control adopts a leader-follower structure (one leader, two followers), and the reconfigurable swarm autonomously drives from the starting point of the building's first floor to the designated location on the second floor according to the following thought process:

(1) After the reconfigurable swarm units complete the system self-check, they start from the starting point in a straight line formation, and execute the leader-following and formation maintenance (unit modality swarm) according to the swarm code embedded in the behavior subtree;

(2) When the reconfigurable swarm turns a corner and the front follower encounters a fault, it switches to the fault recovery under combined modality state, driving the rear follower and the front follower to autonomously complete the combination reconfiguration (unit modality docking);

(3) After reconfiguration, the combined entity regains its ability to behave, continues to turn towards the swarm state with the navigator unit, and carries out coordinated formation tasks (unit and combined hybrid modality swarm);

(4) The reconfigurable swarm completes a layer of detection and turns to encounter a two-layer obstacle in the form of stairs. It then turns to a state of obstacle-crossing reconfiguration, driving the combined entity follower and the navigator in front to complete the combined reconfiguration (combined modality docking);

(5) After the obstacle-crossing reconfiguration, the mobility of the three-unit combined entity is enhanced. The combined entity moves in a straight line in the preset obstacle terrain area with its configuration, until it leaves the area and reaches the second layer position (combined modality movement);

(6) After leaving the obstacle terrain area, the combined entity actively splits into a navigator and a combined entity follower, continuing to turn towards the coordination state (active separation);

(7) The reconfigurable swarm arrives at the second layer preset destination, completing this coordinated detection task (unit and combined hybrid modality swarm) (Fig. 8.5).

The overall trajectory state and velocity commands of the reconfigurable swarm's collaborative process are shown in Figs. 8.6 and 8.7, and the trajectories at each stage

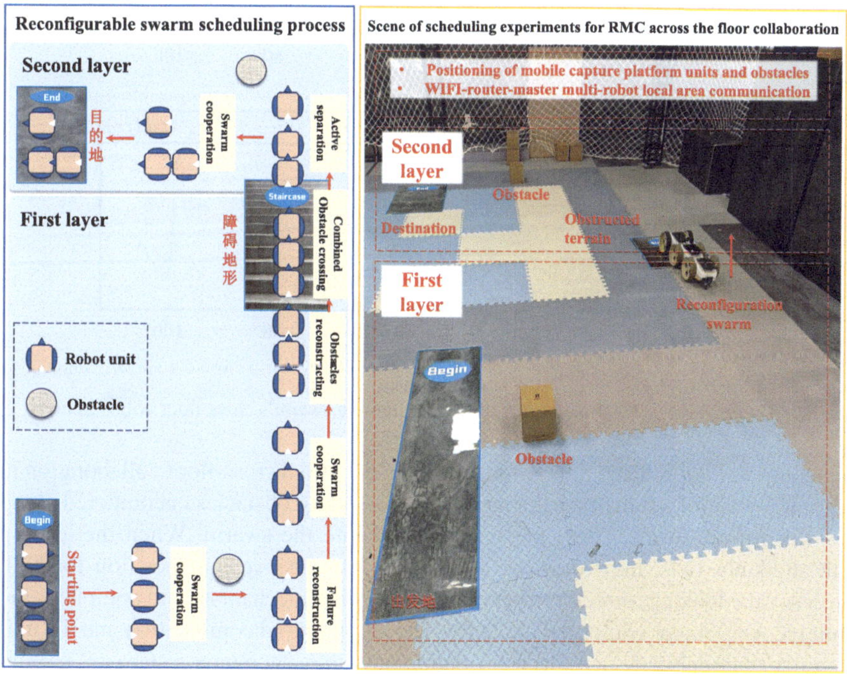

Fig. 8.5 Design and scene setup of scheduling experiments for reconfigurable swarm cross-floor collaboration

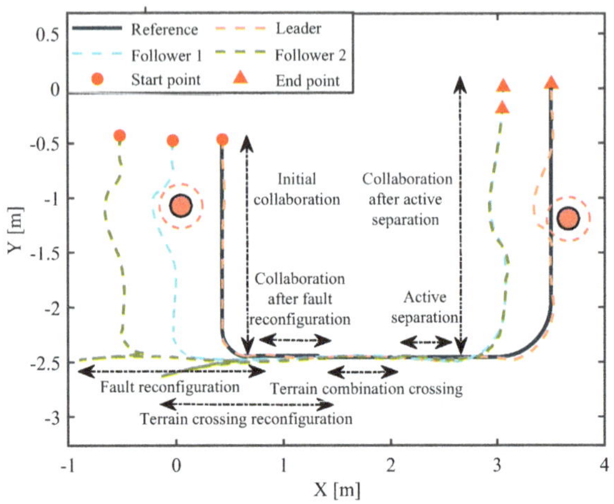

Fig. 8.6 Overall trajectory state of the reconfigurable swarm's cross-floor collaboration

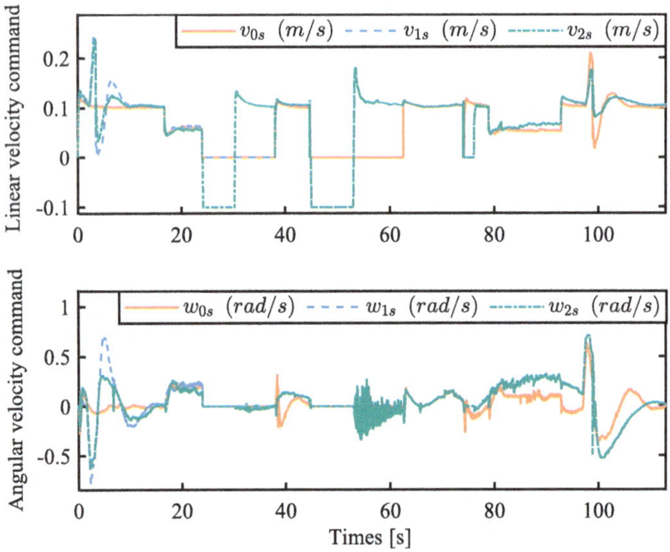

Fig. 8.7 Overall velocity commands for the reconfigurable swarm's cross-floor collaboration

are shown in Figs. 8.8 and 8.9. Three units start the initial cross-floor collaboration in a linear formation on the first floor, smoothly avoiding obstacles encountered along the way thanks to the safety mechanism built into the swarm. When the leading unit suddenly fails, the following unit, due to its inconvenient position for high success rate docking, directly triggers the re-docking mechanism and completes the fault recovery under combined modality. The reconfigured combination and the unit continue to collaborate in a linear formation until they encounter a staircase terrain, which requires crossing the stairs to collaborate on the second floor. The faulty combination and the unit complete the obstacle crossing reconfiguration, and the three-unit combination proceeds until it leaves the staircase terrain. After separation,

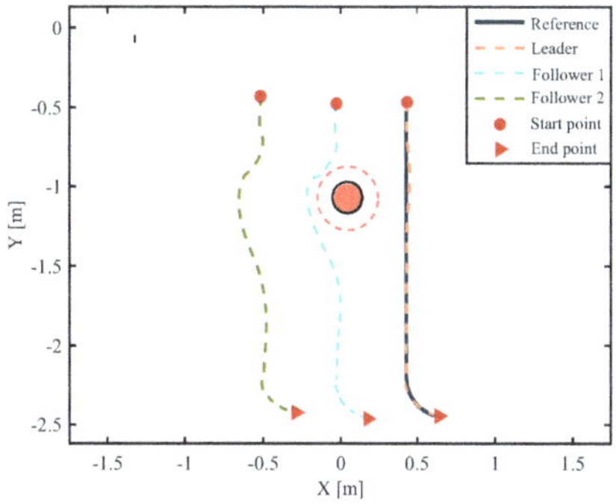

(a) Initial collaboration stage

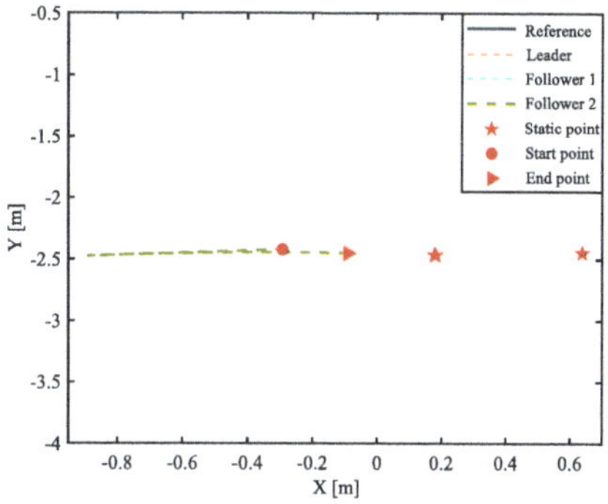

(b) fault recovery under combined modality stage

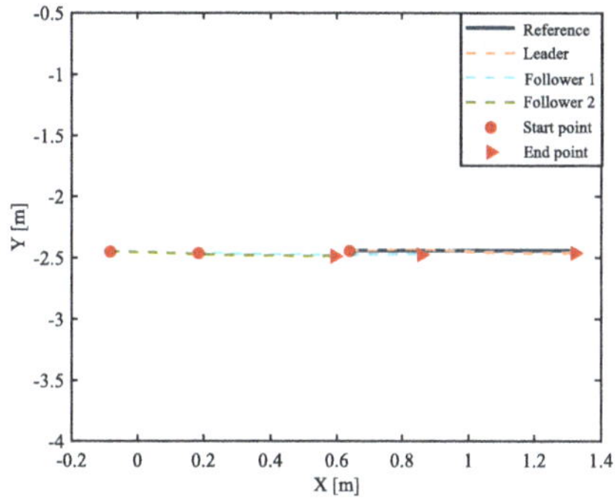

(c) Collaboration stage after fault recovery under combined modality

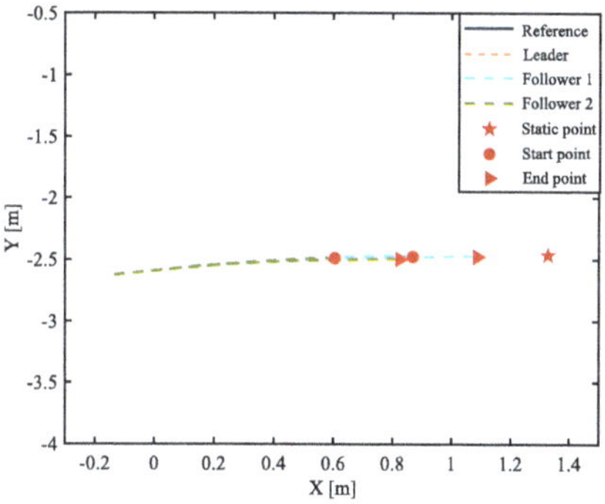

(d) Hurdle reconfiguration stage

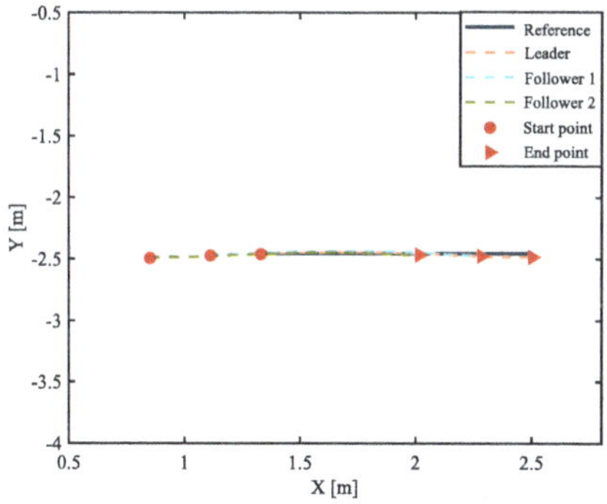

(e) Combination crossing stage

Fig. 8.8 Reconfigurable swarm cross-floor collaboration front stage trajectories

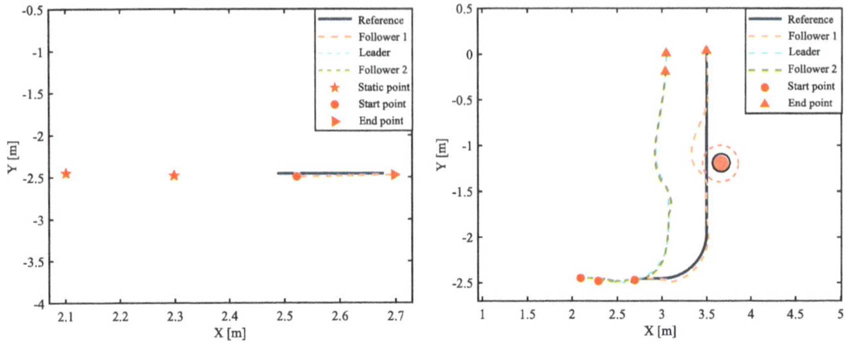

(a) Active separation stage (b) Post-separation coordination stage

Fig. 8.9 Reconfigurable swarm cross-floor collaboration back stage trajectories

they continue to collaborate in a linear formation with the unit and the reconfigured combination. Along the way, the leader encounters an obstacle, but smoothly avoids it thanks to the safety mechanism built into the swarm. Finally, the swarm system reaches the endpoint and stops moving.

(2) Field Terrain Coordination Scene Verification of Four-Unit Reconfigurable Swarm

The outdoor terrain coordination scenario is shown in Fig. 8.10, which emphasizes the flexible convergence and divergence of the reconfigurable swarm under complex terrain tasks, compared to the cross-floor coordination scenario. This scenario introduces a larger unit scale, more obstacle settings, and hazardous terrain. The reconfigurable swarm consists of four units, including one leader and three followers. Four obstacles are set on the coordinated path to verify the safety response during the coordination task. At the same time, the simulation of the ditch and rubble sections of the hazardous terrain verifies the adaptive convergence and divergence of the swarm system during the coordination process. The overall scheduling process first sets four units to perform tasks in a square formation (unit modality swarm), then encounters ditch terrain. The units are oriented according to the scheduling rules and trigger a two-by-two obstacle crossing reconfiguration into a combined modality (unit modality docking), maintaining a straight line formation to continue the obstacle crossing progression (combined modality swarm). After leaving the ditch terrain, they actively separate into unit modality and continue to coordinate in a square formation (unit modality swarm). Immediately after encountering the rubble terrain, the units once again cross obstacles according to the scheduling rules and reconfigure into a combined modality (unit modality docking), and continue the obstacle crossing progression in a straight line formation as a combined modality (combined modality swarm). After leaving the rubble terrain, they actively separate into unit modality and continue to coordinate in a square formation (unit modality swarm) to the final destination, completing the coordination task.

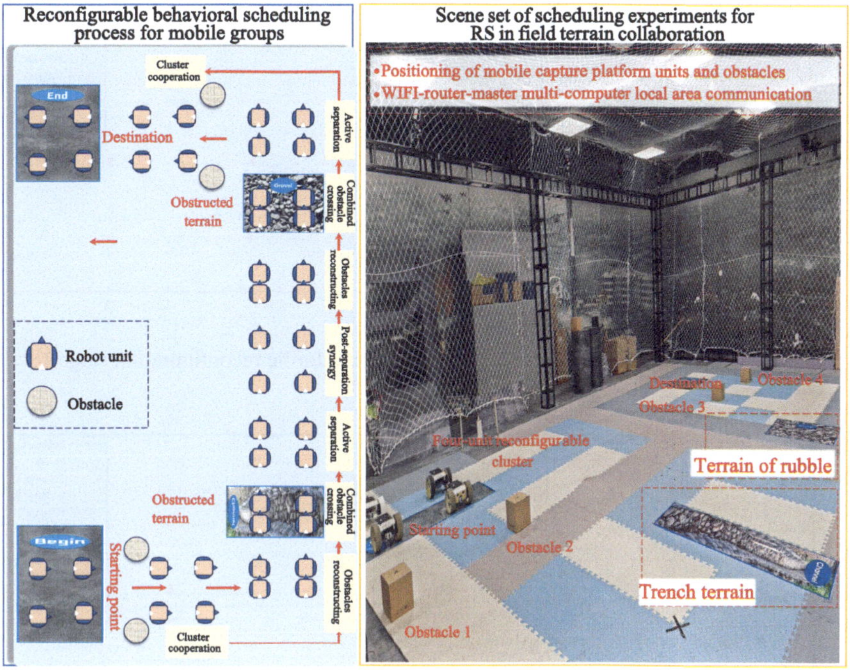

Fig. 8.10 Design and scene construction of reconfigurable swarm outdoor terrain coordination scheduling experiment

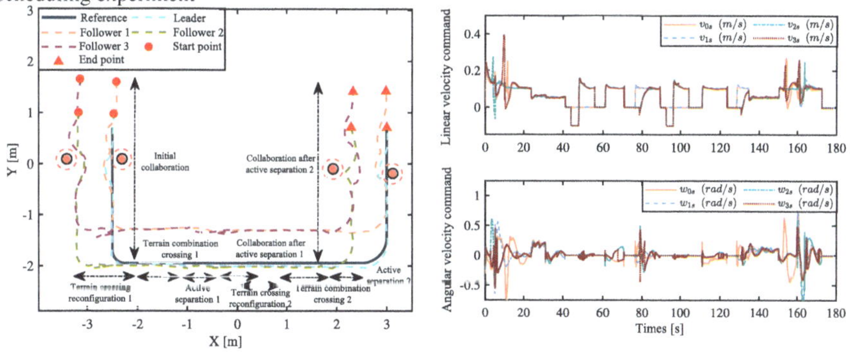

(a) Overall trajectory states (b) Overall velocity commands

Fig. 8.11 Experimental results of reconfigurable swarm outdoor terrain coordination

The overall movement trajectory and velocity commands of the reconfigurable swarm system in outdoor terrain coordination are shown in Fig. 8.11. The phase trajectories of various behaviors such as unit modality coordination, obstacle crossing reconfiguration, combined modality coordination, and active separation are shown in Figs. 8.12 and 8.13. In the task, three stages of unit modality coordination, two stages

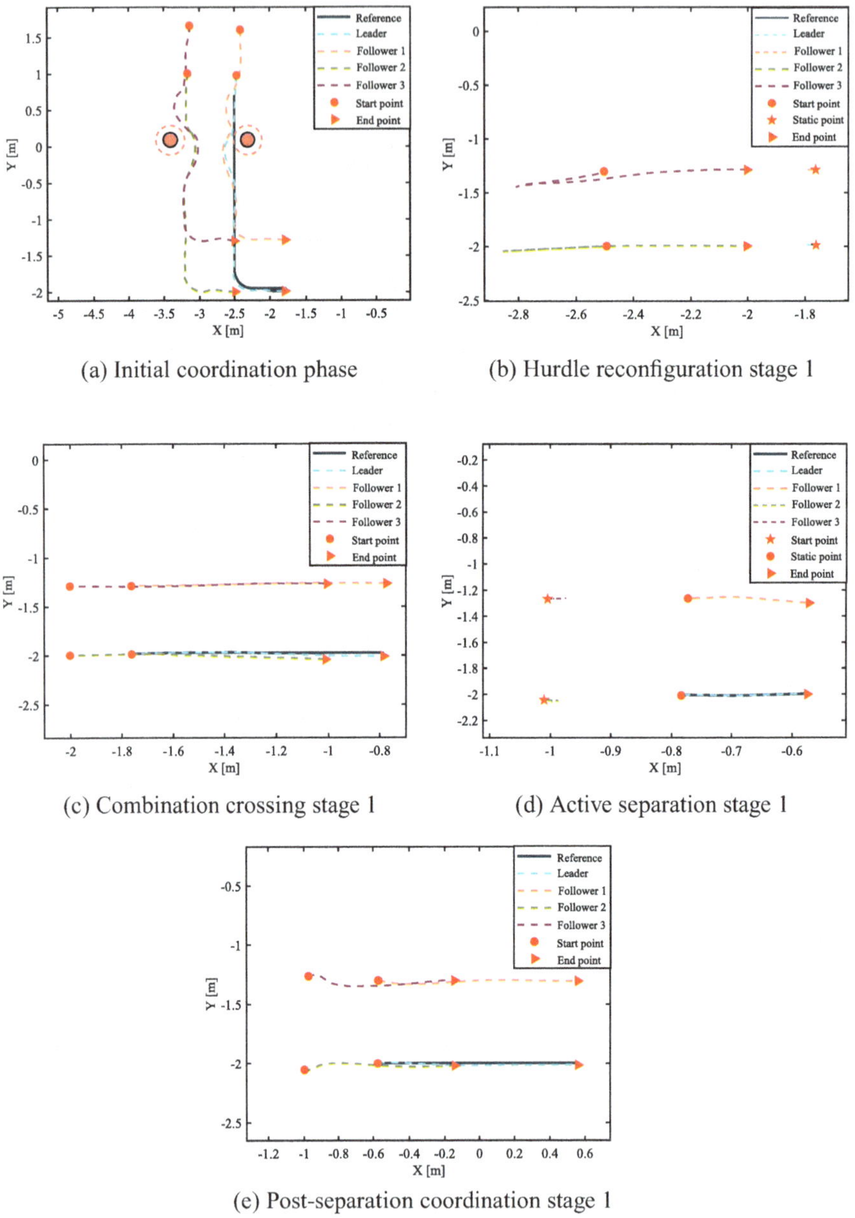

Fig. 8.12 Front stage trajectories in reconfigurable swarm collaboration in wild terrain

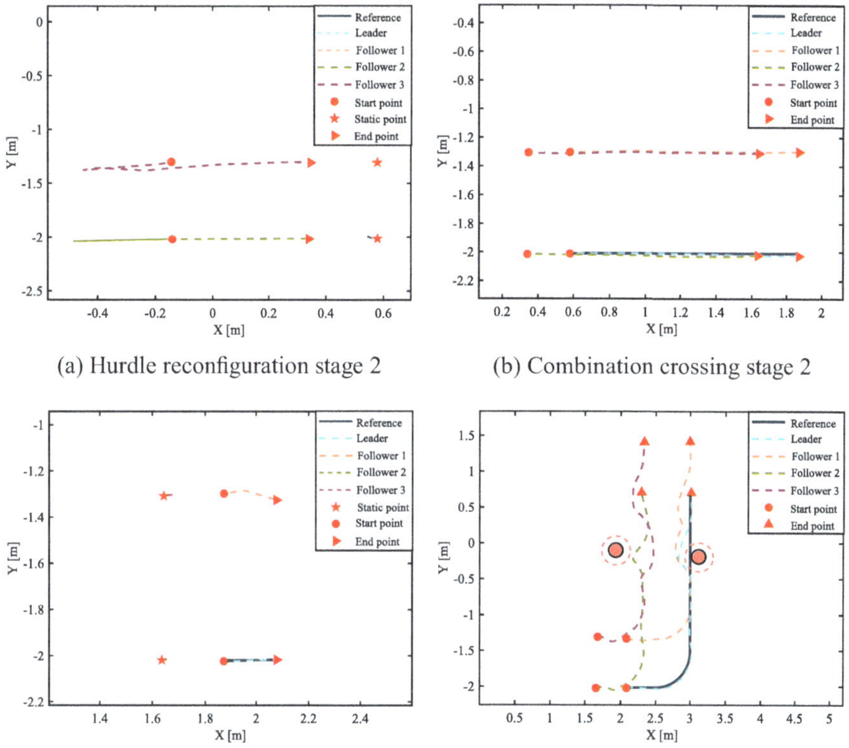

(a) Hurdle reconfiguration stage 2

(b) Combination crossing stage 2

(c) Active separation stage

(d) Post-separation coordination stage 2

Fig. 8.13 Back stage trajectories in reconfigurable swarm collaboration in wild terrain

of obstacle crossing reconfiguration, and two stages of combined modality coordination during obstacle crossing can switch and transition smoothly according to the encountered terrain situation, demonstrating the ability of the reconfigurable swarm system to aggregate and disperse as needed during the overall movement process, and to respond safely when facing multiple obstacles. This scenario is not only more complex in terms of swarm scale and scene setting compared to the cross-floor coordination scenario, but also covers all controller forms in the reconfigurable swarm modality controller set in combination with the cross-floor coordination scenario. It fully demonstrates the ability of the swarm system to balance the aggregation and dispersion needs of the scene in a closed loop, fully verifying the effectiveness of the multimodal behavior scheduling framework and its modality control and modality switching methods proposed in this chapter, as well as the adaptability advantages and application potential of the multi-modal motion control scheme of the reconfigurable swarm system studied in this book in future real-world operations.

8.4 Summary

Studying multimodal behavior scheduling is a crucial step in multimodal transition and the ultimate task loop of multimodal motion control. This chapter not only uses finite state machines and behavior tree decision-making methods to design a behavior scheduling framework for smooth multimodal transition of reconfigurable swarm systems, but also combines the theoretical methods of intra-modal control and inter-modal switching from Chaps. 4–7 to further design the modular unit/combined modality reconfiguration transformation, and the hybrid swarm of modular unit/combined modality before and after reconfiguration in the action nodes of the behavior tree. Finally, using two specific cross-terrain task scenarios of reconfigurable swarm cross-floor coordination and outdoor terrain coordination, we fully verify the reconfiguration transformation of each modality and its swarm under the multimodal overall operation control framework, the convergence and divergence of the swarm system's capabilities as needed, and highlight the system's enhanced adaptability in complex scenarios, as well as its advantages in balancing "passability" and "execution efficiency".

References

1. Motes, J., Sandstrom, R., Lee, H., Thomas, S., Amato, N.M.: Multi-robot task and motion planning with subtask dependencies. IEEE Robot. Autom. Lett. **5**(2), 3338–3345 (2020)
2. Chen, Y., Rosolia, U., Ames, A.D.: Decentralized task and path planning for multi-robot systems. IEEE Robot. Autom. Lett. **6**(3), 4337–4344 (2021)
3. Bolu, A., Korcak, O.: Adaptive task planning for multi-robot smart warehouse. IEEE Access **9**, 162650–162667 (2021)
4. Booth, K.E.C., Tran, T.T., Nejat, G., Beck, J.C.: Mixed-integer and constraint programming techniques for mobile robot task planning. IEEE Robot. Autom. Lett. **1**(1), 500–507 (2016)
5. Touzani, H., Hadj-Abdelkader, H., Seguy, N., Bouchafa, S.: Multi-robot task sequencing and automatic path planning for cycle time optimization: Application for car production line. IEEE Robot. Autom. Lett. **6**(2), 1335–1342 (2021)
6. Jose, K., Pratihar, D.K.: Task allocation and collision-free path planning of centralized multi-robots system for industrial plant inspection using heuristic methods. Robot. Auton. Syst. **80**, 34–42 (2016)
7. Lashkari, N., Biglarbegian, M., Yang, S.X.: Development of a novel robust control method for formation of heterogeneous multiple mobile robots with autonomous docking capability. IEEE Trans. Autom. Sci. Eng. **17**(4), 1759–1776 (2020)
8. Jing, G., Tosun, T., Yim, M., Kress-Gazit, H.: An end-to-end system for accomplishing tasks with modular robots: perspectives for the AI community. In: Proceedings of 26th International Joint Conference on Artificial Intelligence (IJCAI), pp. 4879–4883 (2017)

The manufacturer's authorised representative in the EU is Springer Nature Customer Service Centre GmbH, Europaplatz 3, 69115 Heidelberg, Germany. If you have any concerns regarding our products, please contact ProductSafety@springernature.com

Printed and bound by CPI Group (UK) Ltd, Croydon, CR0 4YY
01/10/2026
02225921-0001